1 MONTH OF FREE READING

at
www.ForgottenBooks.com

By purchasing this book you are eligible for one month membership to ForgottenBooks.com, giving you unlimited access to our entire collection of over 1,000,000 titles via our web site and mobile apps.

To claim your free month visit:
www.forgottenbooks.com/free906160

* Offer is valid for 45 days from date of purchase. Terms and conditions apply.

ISBN 978-0-265-89636-5
PIBN 10906160

This book is a reproduction of an important historical work. Forgotten Books uses state-of-the-art technology to digitally reconstruct the work, preserving the original format whilst repairing imperfections present in the aged copy. In rare cases, an imperfection in the original, such as a blemish or missing page, may be replicated in our edition. We do, however, repair the vast majority of imperfections successfully; any imperfections that remain are intentionally left to preserve the state of such historical works.

Forgotten Books is a registered trademark of FB &c Ltd.
Copyright © 2018 FB &c Ltd.
FB &c Ltd, Dalton House, 60 Windsor Avenue, London, SW19 2RR.
Company number 08720141. Registered in England and Wales.

For support please visit www.forgottenbooks.com

UNITED STATES DEPARTMENT OF AGRICULTURE
BULLETIN No. 862

Contribution from the Bureau of Biological Survey
E. W. NELSON, Chief

FOOD HABITS OF SEVEN SPECIES OF AMERICAN SHOAL-WATER DUCKS

By

DOUGLAS C. MABBOTT, Assistant in Economic Ornithology

CONTENTS

	Page		Page
Introduction	1	Blue-winged Teal	22
Gadwall	2	Cinnamon Teal	28
Baldpate	10	Pintail	31
European Widgeon	16	Wood Duck	37
Green-winged Teal	17		

UNITED STATES DEPARTMENT OF AGRICULTURE

 # BULLETIN No. 862

Contribution from the Bureau of Biological Survey
E. W. NELSON, Chief

Washington, D. C.   PROFESSIONAL PAPER.   December 30, 1920

FOOD HABITS OF SEVEN SPECIES OF AMERICAN SHOAL-WATER DUCKS.

By Douglas C. Mabbott,[1] *Assistant in Economic Ornithology.*

CONTENTS.

	Page.		Page.
Introduction	1	Blue-winged teal	22
Gadwall	2	Cinnamon teal	28
Baldpate	10	Pintail	31
European widgeon	16	Wood duck	37
Green-winged teal	17		

INTRODUCTION.

The wild ducks of the United States belong to three main groups: The mergansers (Merginae), known also as fish ducks or sawbills; the river ducks (Anatinae), called also shoal-water, puddle, plash, or tipping ducks; and the sea ducks (Fuligulinae), otherwise known as deep-water or diving ducks. This bulletin treats of the food habits of eight species [2] of shoal-water ducks, one of which, the European widgeon, is only a straggler in the United States. Wild ducks are our most important game birds, their value to the people of the

---

[1] Douglas Clifford Mabbott, author of this bulletin, was a member of the heroic Sixth Regiment, United States Marine Corps, and participated in all the hard fighting done by that organization at Bouresches, Belleau Wood, Soissons, and in the St. Mihiel salient. He was killed in action September 15, 1918, while taking part in an advance in the battle of St. Mihiel, and was buried near the village of Xammie, near Thiaucourt, France. He was born at Arena, Wis., March 12, 1893, and became a member of the staff of the Biological Survey, December 1, 1915.—EDITOR.

[2] Three other species, the mallard, black duck, and southern black duck, are treated in Bull. 720, U. S. Dept. Agr., Food Habits of the Mallard Ducks of the United States, by W. L. McAtee, pp. 35, pl. 1, Dec. 23, 1918.

NOTE.—This bulletin presents a technical study of the food habits of seven species of American shoal-water ducks: The gadwall, the baldpate, the green-winged, blue-winged, and cinnamon teals, the pintail and the wood duck; and includes a brief note on the European widgeon, which is a straggler in the United States. The vegetable food preferences exhibited will serve as guide to certain wild-duck foods that may be propagated when it is sought to increase the numbers of these valuable game ducks either in the wild state or in domestication. For specific information on this topic, see Bull. 205, U. S. Dept. Agr., Eleven Important Wild-duck Foods, in which are discussed musk grass, duckweeds, frogbit, thalia, water elm, swamp privet, eelgrass, widgeon grass, watercress, waterweed, and coontail; pp. 25, figs. 23, May 20, 1915; also Bull. 465, Propagation of Wild-duck Foods, in which are discussed wild rice, wild celery, pondweeds, delta potato, wapato, chufa, wild millet, and banana waterlily; pp. 40, figs. 35, Feb. 23, 1917.

179375°—20——1

United States totaling hundreds of thousands of dollars. Some of the species covered by this bulletin are among the most valuable, as the pintail, gadwall, baldpate, and green-winged teal.

The ducks here discussed have not thus far been utilized in duck farming to so great an extent as the mallard and black ducks, but the wood duck and the green-winged teal have proved to be adapted to such use, and possibly further experiments will result as successfully with some of the other species. Information presented in the following pages shows the preferences of these ducks among vegetable foods, matters which should be heeded in attempting to establish the more or less natural conditions which probably will be found necessary for success in the propagation of some of the species in inclosures.

GADWALL.

(*Chaulelasmus streperus*.)

PLATE I.

The gadwall, or gray duck, as it is sometimes called, is almost cosmopolitan in its distribution, breeding commonly in Europe, Asia, and North America and ranging south in winter to southern Asia, some distance into Africa, and in North America to the southern end of Lower California and to southern Puebla. East of the Mississippi River, however, and north of North Carolina, the bird is rare, and in New England is found only as a straggler. It breeds in most of the western United States and in southern Canada, but its principal breeding range for North America is in the prairie district extending from Manitoba and western Minnesota to the Rocky Mountains, south to Nebraska, and north to Saskatchewan.

The adult male gadwall is distinguished particularly by the scale-like markings on the breast, each feather on the lower neck and breast being black with a white crescent and a white border, producing a peculiar mottled or barred effect. The bird has a prominent white speculum or wing patch, bordered in front by black, with an area of chestnut-brown on the forepart of the wing, comprising the middle wing coverts. The remainder of the plumage is chiefly gray or brownish, streaked with black. The female lacks the chestnut wing coverts, and the breast and sides are buffy with the barred appearance less distinct.

FOOD HABITS.

In habits the gadwall resembles the mallard, feeding either on dry land or in shallow water near the edges of ponds, lakes, and streams, where it gets its food by "tilting" or standing on its head in the water. The food of both the gadwall and the baldpate, however, is quite different in some respects from that of the mallard. These two feed to a very large extent upon the leaves and stems of water

GADWALL (CHAULELASMUS STREPERUS).
Male on right; female on left.

plants, paying less attention to the seeds, while the mallard feeds indiscriminately on both or even shows some preference for the seeds. In fact, in respect to the quantity of foliage taken, the gadwall and the baldpate are different from all other ducks thus far examined by the Biological Survey. They are also more purely vegetarian, their diet including a smaller percentage of animal matter than that of any of the other ducks.

For a determination of the food habits of the gadwall 417 stomachs[3] were available. These were from 19 States and Canada, and their collection extended over a period of 31 years. Only 24 were taken during the five months from April to August and their contents were not included in computing the average percentages, so that the results thus obtained apply only to the fall and winter months.

Considerably more animal food is taken in summer than in winter, owing, of course, to the fact that more is available at that time of year. The percentage of animal food for the summer months is higher also because there are included in the averages analyses of numerous stomach contents of ducklings, which feed to a great extent upon insects. All of the 11 stomachs collected during the month of July (9 from North Dakota and 2 from Utah) were of young ducklings. A computation of the average contents of this series produced the following results: Water bugs, 56.18 per cent; beetles, 7.09; flies and their larvae, 2; nymphs of dragonflies and damselflies, 0.27; other insects, 2; total animal food, 67.54 per cent. Pondweeds, 12.55 per cent; grasses, 5.09; sedges, 2; water milfoils, 0.55; smartweeds, 0.09; miscellaneous, 12.18; total vegetable food, 32.46 per cent.

Of the remaining 13 stomachs collected in summer, all but two were from mature birds. Their contents averaged 11.17 per cent animal food and 88.83 per cent vegetable; 5.28 per cent, or nearly half the animal food, consisted of snails. Thus it will be seen that, so far as can be judged from the contents of such a limited number of stomachs, the summer food of the adult birds does not differ greatly from their winter food.

A rather large proportion of the total number of stomachs (131) was from birds taken in Louisiana. These furnished the bulk of the collections for November, February, and March, but averaged much the same as those from the other States, the principal items consisting of sedges, pondweeds, *Sagittaria* tubers, grasses, some cultivated rice, and mollusks. Arkansas contributed 57 stomachs; Utah, 53; North Carolina, 30; North Dakota, 22; and Florida, 20; the remainder being scattered.

---

[3] Seventy-six of these were examined by W. L. McAtee.

VEGETABLE FOOD.

As computed from the contents of 362 stomachs collected during the six months from September to March, 97.85 per cent of the food of the gadwall consists of vegetable matter. This is made up as follows: Pondweeds, 42.33 per cent; sedges, 19.91; algae, 10.41; coontail, 7.82; grasses, 7.59; arrowheads, 3.25; rice and other cultivated grain, 1.31; duckweeds, 0.61; smartweeds, 0.59; wild celery and waterweed, 0.53; waterlilies, 0.52; madder family, 0.37; and miscellaneous, 2.61 per cent.

PONDWEEDS (NAIADACEAE), 42.33 PER CENT.

Of the 417 gadwalls whose stomachs were examined, 155 had eaten true pondweeds (*Potamogeton* spp.), 112 widgeon grass (*Ruppia maritima*), 20 horned pondweed (*Zannichellia palustris*), 17 bushy pondweed (*Najas flexilis*), 3 eelgrass (*Zostera marina*), and 8 pondweeds which were too far advanced in the process of digestion to be further identified. In nearly all cases the pondweed food consisted chiefly of leaves and stems, and sometimes buds and tubers. Seeds were often present, sometimes in considerable numbers, but as a rule they appeared to be merely incidental. Pondweeds are undoubtedly the favorite food of this species, as well as of the baldpate, and they are eaten very greedily. The gullet of one gadwall taken in Texas in November contained a mass of the foliage of small pondweed (*Potamogeton pusillus*) the size of a billiard ball. A series of 26 stomachs taken in North Carolina in December contained practically nothing but the leaves and stems of pondweeds, including true pondweeds, bushy pondweed, and widgeon grass. Many of these stomachs were crammed. Often a few of the seeds were present, and three stomachs contained in addition a few sedge seeds. Other rather large series of gizzards containing chiefly foliage of pondweeds were taken in Florida, Louisiana, Utah, and North Dakota.

SEDGES (CYPERACEAE), 19.91 PER CENT.

The sedges, second in favor among the food items of the gadwall, constitute an important exception to this bird's rule of feeding upon the leaves and stems of plants rather than upon the seeds, for the leaves and stems of practically all the sedges are coarse, fibrous, or even woody, and do not make choice morsels. On the other hand, the seeds are a favorite item of food among most fresh-water ducks. The sedge seeds most often eaten by the gadwall were those of three-square (*Scirpus americanus*), by 150 birds; prairie bulrush (*S. paludosus*), by 27; salt-marsh bulrush (*S. robustus*), by 24; unidentified bulrushes (*Scirpus* spp.), by 47; saw grass (*Cladium effusum*), by 68; and chufas (*Cyperus* spp.), by 31. A considerable number of birds from the Mississippi Delta, Louisiana, had been feeding during

January and February almost exclusively on the seeds of three-square. Some had eaten also the rootstocks of bulrushes, probably of the same species as the seeds; others from the same general region had varied their diet by feeding to some extent upon the delta potato (tubers of *Sagittaria platyphylla*), and a few snails. Bulrush seeds, however, usually constituted the bulk of the stomach contents. Several gizzards contained no fewer than 1,800 to 3,000 seeds.

ALGAE, 10.41 PER CENT.

It is not surprising that in a duck which feeds so freely upon the foliage of aquatic vegetation, algae formed more than one-tenth of the total stomach contents. These were eaten most freely in spring, the maximum consumption being 21.67 per cent of the total food for the month of March, and the minimum, 1.64 per cent in December. Most of the algae eaten consisted of musk grass (*Chara* spp.), but several other kinds were present.

COONTAIL (*Ceratophyllum demersum*), 7.82 PER CENT.

So far as known, the gadwall is the only duck which feeds to any extent upon the foliage of coontail, which gets its common name from a fancied resemblance in the shape of its finely branching stems and leaves to the tail of a raccoon. It is also called hornwort, hornweed, and morassweed. Many other species of ducks commonly feed upon the hard, horny coated seeds of the plant, but a series of 50 gadwalls taken in December, 1909, along the Mississippi River in northwestern Arkansas, had eaten large quantities of the leaves and tips of the stems, many to the exclusion of all other food.

The contents of these 50 stomachs averaged as follows: Coontail, 87.72 per cent; duckweeds, 3.88; seeds of buttonbush, 1.66; pondweeds, 1.6; algae, 1; sedges, 0.16; miscellaneous vegetable matter, 3.24; statoblasts of fresh-water bryozoa, 0.6; and water bugs, 0.14 per cent. It is possible that if stomachs of the baldpate had been available from the same region, this bird also might have shown a taste for the foliage of coontail. However, three other gadwall stomachs (one from Colorado and two from Louisiana) contained considerable quantities of the plant, while only one of the entire collection of baldpates had eaten it to an appreciable extent.

GRASSES (GRAMINEAE), 7.59 PER CENT; AND CULTIVATED GRAINS, 1.31 PER CENT.

Considerable quantities of grass were found in stomachs collected during the spring months, especially March, when the tender young shoots are plentiful throughout the greater part of the ducks' winter range. This consisted largely of the shoots and young leaves of switchgrass (*Panicum repens* and others of the same genus), but there were also present meadow grass (*Poa* sp.), saltgrass (*Distichlis spicata*), little barley (*Hordeum pusillum*), crab grass (*Syntherisma sanguinalis*), wild millet (*Echinochloa crus-galli*), foxtail (*Chaetochloa glauca*), cutgrass (*Zizaniopsis miliacea*), rice cut-grass (*Homalocenchrus oryzoides*), salt-marsh grass (*Spartina* sp.), manna grass (*Panicularia* sp.), *Monanthochloë littoralis*, and a few others, unidentified. Several of these were represented only by the seeds, and then they usually constituted a comparatively small part of the stomach contents.

The cultivated grain was tabulated separately from the remainder of the grasses because of the economic interest attached to it. It consisted, however, almost entirely of rice found in the gizzards of several Louisiana birds taken in February, and was undoubtedly waste grain. One stomach taken in Oregon in January was crammed with grains of barley; and another, from North Carolina in February, contained several kernels of corn. Obviously these also were of no economic importance. The rice, barley, and corn together amounted to 1.31 per cent of the contents of the whole number of stomachs.

WATER PLANTAIN FAMILY (ALISMACEAE). 3.25 PER CENT.

One of the favorite items of food among many species of ducks in the lower Mississippi Valley during the fall and winter months is the delta potato, as the starchy tubers of a species of arrowhead (*Sagittaria platyphylla*) are called.[4] These constitute an especially important food item among ducks wintering on the Mississippi Delta, Louisiana, where the tubers grow in great abundance and the variety of duck food is not great. Many gadwall stomachs from this region contained only three items of food, which also have been found to be the typical diet of several other species when wintering on the Delta: these were the seeds of three-square (*Scirpus americanus*), the delta potato, and a species of snail (*Neritina virginea*), very abundant there. The stomach contents of a series of 27 gadwalls taken near the end of the Delta in November averaged as follows: Seeds of three-square (with a few of salt-marsh bulrush), 44.55 per cent; delta potato, 20.89; pondweeds, 13.78; and snails, 7.11 per cent; several minor items, as algae, coontail, duckweeds, and a few insects made up the remainder.

DUCKWEEDS (LEMNACEAE), 0.61 PER CENT.

It is rather surprising that a duck which shows such a marked preference for the foliage of aquatic vegetation as the gadwall should not have eaten duckweeds to a greater extent. These are small floating plants, often present in such abundance in ponds, lakes, and sluggish streams as completely to cover large areas of their surfaces. The little plants are luscious and tender, and afford a favorite article of food for many species of duck. Large numbers of the gadwall

---

[4] Bull. 465, U. S. Dept. Agr., pp. 21–24, 1917.

stomachs examined were collected in the swamps of Louisiana, Arkansas, and other localities where duckweeds abound, but the majority failed to disclose any duckweeds. Only 17 of the total number of ducks had eaten duckweeds (*Lemna* spp.), and some of these only in very limited quantities.

SMARTWEEDS (POLYGONACEAE), 0.59 PER CENT.

The Polygonaceae is one of the families of plants of which the seeds alone furnish an important article of food for birds. This very probably is the reason why smartweeds are only one of the minor items in the food of the gadwall. The following species were identified from the stomachs examined: Dock-leaved smartweed (*Polygonum lapathifolium*), found in 5 stomachs; water smartweed (*P. amphibium*), in 3; and knotweed (*P. aviculare*), Pennsylvania smartweed (*P. pennsylvanicum*), water pepper (*P. hydropiper*), lady's-thumb (*P. persicaria*), mild water pepper (*P. hydropiperoides*), and prickly smartweed (*P. sagittatum*), in 2 each. Seeds of black bindweed (*Polygonum convolvulus*) and another species (*P. opelousanum*) were present in 1 each, unidentified smartweeds in 2, and seeds of dock (*Rumex* spp.) in 2. Smartweed seeds were present usually in small numbers, but the gullet of one bird taken in Montana was crammed with about 3,000 seeds of water pepper, in addition to a few of dock-leaved smartweed.

FROGBIT FAMILY (HYDROCHARITACEAE), 0.53 PER CENT.

Wild celery (*Vallisneria spiralis*) was found in the stomachs of 3 birds shot on Mobile Bay, Alabama, and waterweed (*Philotria* spp.) had been eaten in generous quantity by a bird from southern Wisconsin. Wild celery is a very important food item of some species of ducks.

WATERLILY FAMILY (NYMPHAEACEAE), 0.52 PER CENT.

Two gadwall stomachs collected in Florida were filled with the seeds of a white waterlily (*Castalia* sp.), one containing about 1,100 and the other 1,200 seeds. Another from the same State contained 28 of the hard ovoid seeds of water shield (*Brasenia schreberi*).

MADDER FAMILY (RUBIACEAE), 0.37 PER CENT.

Twenty-three gadwalls had eaten seeds of buttonbush (*Cephalanthus occidentalis*). These seeds are narrowly wedge shape and are borne like miniature sycamore balls in spherical clusters on the ends of the branches of the plant, which is a shrub or small tree growing in wet places. They had been eaten by few of the ducks in any great numbers, but in some instances they constituted the greater part of the stomach contents.

MISCELLANEOUS VEGETABLE FOOD, 2.61 PER CENT.

A large number of miscellaneous items made up the remainder of the gadwall's vegetable food. The stomach of one duck from the mouth of Bear River, Utah, was filled with remains of the stems, leaves, and seeds of picklegrass (*Salicornia ambigua*). A young duck from the same region had made a meal of willow catkins (*Salix* sp.). Several gizzards from the wooded swamps of Arkansas contained fragments of scales from the cones of bald cypress (*Taxodium distichum*), and one was entirely filled with galls from cypress leaves. Many from this region also contained the seeds, or fragments of seeds, of grapes (*Vitis* sp.), hackberry (*Celtis* sp.), holly (*Ilex* sp.), and sumachs (*Rhus* spp.). Seeds of beggar ticks, or bur marigold (*Bidens* sp.), water milfoil (*Myriophyllum* sp.), bottle brush (*Hippuris vulgaris*), crowfoot (*Ranunculus* sp.), water pennywort (*Hydrocotyle* sp.), dodder (*Cuscuta* sp.), myrtle (*Myrica* sp.), bur reed (*Sparganium* sp.), heliotrope (*Heliotropium indicum*), and many others, eaten in small quantities, completed the vegetable food of the species.

ANIMAL FOOD.

As has been stated previously, the proportion of animal food taken by the gadwall is very small, amounting to only 2.15 per cent of the contents of the stomachs examined, exclusive of the few scattered items taken during the months from April to August. In these the presence of several stomachs of ducklings caused the average percentage of animal food to run considerably higher. The figures given were compiled from the contents of the 362 stomachs collected during the fall and winter months, from September to March.

MOLLUSKS (MOLLUSCA), 1.6 PER CENT.

About three-fourths of the animal food of the gadwall, or 1.6 per cent of the total, consisted of mollusks. In 6 April stomachs (not included in this average) they amounted to 15.83 per cent of the monthly food. In the fall and winter months they ranged from nothing in September to 4 per cent in January. Eight species of snails were identified, while there were unidentified fragments of snails in 5 stomachs and unidentified bivalves in 3. The most important snail was *Neritina virginea*, which is very common on the Mississippi Delta and constitutes one of the principal items of food of many species of ducks wintering in that region. This had been eaten by 25 gadwalls and ranged from a mere trace to 70 per cent of the food present.

INSECTS (INSECTA), 0.39 PER CENT.

Insects amounted to only 0.39 per cent of the total food. These consisted of caddisflies and their larvae (Phryganoidea), 0.19 per cent; flies and their larvae (Diptera), 0.07; bugs (Hemiptera), 0.05;

beetles (Coleoptera), 0.04; dragonflies and damselflies and their nymphs (Odonata), 0.01; and other insects, 0.03 per cent.

One Oregon bird had made almost a full meal of adult caddisflies in October, and the tube-shaped larval cases were found in the stomachs of 8 others.

The Diptera usually consisted of larvae or pupae, but occasionally of adult flies. Six families were represented, as follows: Craneflies (Tipulidae), found in 1 stomach; midges (Chironomidae), in 10; soldierflies (Stratiomyidae), in 2; horseflies (Tabanidae), in 1; Borboridae, in 3; and Ephydridae, in 8.

The bugs taken were chiefly aquatic. Water boatmen (Corixidae) had been eaten by 25 gadwalls, creeping water bugs (Naucoridae) by 6, and water striders (Gerridae) by 4, while shorebugs (Saldidae), stink bugs (Pentatomidae), and plant hoppers (Fulgoridae) were taken by 1 each.

The most common Coleoptera were water scavenger beetles (Hydrophilidae), predacious diving beetles (Dytiscidae), ground beetles (Carabidae), leaf beetles (Chrysomelidae), and weevils (Rhynchophora). Other families represented were rove beetles (Staphylinidae), larder beetles (Dermestidae), ladybugs (Coccinellidae), pill beetles (Byrrhidae), leaf chafers (Scarabaeidae), darkling beetles (Tenebrionidae), flower beetles (Anthicidae), and blister beetles (Meloidae). Of the 362 birds taken during the fall and winter months, only 23 had eaten beetles, and these never amounted to more than 4 per cent of the stomach contents. Of 11 ducklings taken in July, however, all but one had eaten beetles; in three instances these amounted to 15 per cent, and constituted 7.09 per cent of the food of all.

Two gadwalls had eaten nymphs of dragonflies (Anisoptera), two those of damselflies (Zygoptera), and one an odonate nymph, too badly ground to be identified.

The miscellaneous insects consisted of a few ants, ichneumons, etc. (Hymenoptera), and a caterpillar (Lepidoptera). Together they amounted to only 0.03 per cent.

CRUSTACEANS (CRUSTACEA), 0.08 PER CENT.

Crustaceans evidently are not much sought after by the gadwall. Twenty-one birds had eaten very small bivalved crustaceans (Ostracoda), usually in small numbers. Three gizzards contained the fingers of crabs, two the remains of crawfish, and one a sowbug (*Oniscus asellus*). Altogether, crustaceans furnished only 0.08 per cent of the gadwall's food.

MISCELLANEOUS ANIMAL FOOD, 0.08 PER CENT.

The stomach of a gadwall from an open lake in northeastern Arkansas contained several hundred of the small reproductive buds, or statoblasts, of fresh-water Bryozoa. These are simple animal organisms which grow in colonies resembling masses of jelly, attached to submerged brush. Bits of hydroids (animals closely related to the corals) were found in 2 stomachs; spiders, in 3; water mites (Hydrachnidae), in 3; and the teeth or scales of small fish, in 2.

BALDPATE.

Mareca americana.

PLATE II.

Roughly speaking, the range of the baldpate, or American widgeon, includes practically all of North America. Its breeding range extends from Lake Michigan and Hudson Bay west to the Pacific Ocean and from Wisconsin, Colorado, and Oregon north to central Alaska, the Mackenzie Valley, and Fort Churchill. It does not breed commonly, however, east of Minnesota or south of North Dakota. Along the Atlantic Coast it is common in migration as far as Chesapeake Bay, and is only a straggler in New England and eastern Canada. In winter it is found as far south as Florida, Cuba, and Guatemala, and rarely in Costa Rica, Jamaica, Porto Rico, and Trinidad. Many individuals winter as far north as southern British Columbia, Utah, New Mexico, Illinois, and Chesapeake Bay, and a few occasionally remain in southern New England.

The adult baldpate is distinguished by the following characters: There is a large area of white on the wings in front of the speculum, which is black with a narrow green area near its front edge; the top of the head, including the forehead, is white, producing the bald appearance which gives the bird its name. Just below the "bald spot," covering each side of the head from the eye back to and including the nape of the neck, is a broad stripe of glossy green; below this the head and neck are mottled gray, the upper breast and sides pinkish brown, lower breast and belly white, under tail-coverts and outer upper tail-coverts black; the back is finely barred with black and gray or buff, and the rump is mostly white. The female lacks the white crown and green headband; the back is more coarsely mottled and streaked, and the white of the wings is less prominent.

FOOD HABITS.

The feeding habits of the baldpate are in general very similar to those of the gadwall. In some respects the similarity of the results obtained by computing the average percentages of certain elements of food in a large number of stomachs of each species is quite remarkable. For instance, the average proportion of pondweeds (Naiadaceae) found in the gadwall stomachs was 42.33 per cent, while in the case of the baldpate it was 42.82 per cent. There are a few slight differences in the food habits of the two species, however. The bald-

BALDPATE (MARECA AMERICANA).
Male on right; female on left.

pate appears to be even less of a seed eater than the gadwall. Sedges (Cyperaceae), consisting almost entirely of seeds, amounted to 19.91 per cent of the food of the gadwall, but to only 7.41 per cent of the food of the baldpate. The baldpate also ate more wild celery (*Vallisneria spiralis*), grasses, and water milfoils (*Hippuris vulgaris* and *Myriophyllum* sp.), but much less coontail (*Ceratophyllum demersum*).

Investigation of the food habits of the baldpate consisted chiefly of an examination of the contents of 255 stomachs,[5] collected (all but 4) during the months from September to April, inclusive, from 25 States, 4 Canadian Provinces, Alaska, and Mexico. With the exception of series of 53 from Utah, 50 from Oregon, and 29 from North Carolina, they were very evenly distributed in numbers among the different States and Provinces. Four stomachs of birds shot in May and June, together with 22 others which were too nearly empty to allow accurate estimates of percentages of food contents, were not included in the computation, so that the results given are from the remaining 229 stomachs. In the list of food items, however, material from all stomachs is included.

VEGETABLE FOOD.

The vegetable food of the baldpate for the 8 months from September to April averaged 93.23 per cent. This consisted of the following items in the order of their importance: Pondweeds, 42.82 per cent; grasses, 13.9; algae, 7.71; sedges, 7.41; wild celery and waterweed, 5.75; water milfoils, 3.48; duckweeds, 2.2; smartweeds, 1.47; arrow-grass, 0.36; waterlilies, 0.26; coontail, 0.24; and miscellaneous, 7.63 per cent.

PONDWEEDS (NAIADACEAE), 42.82 PER CENT.

Pondweeds are by far the most important item of food of the baldpate, as well as of the gadwall and several other species of ducks. Of the 229 baldpate stomachs, 157, or more than two-thirds, contained pondweeds in some form or other. True pondweeds (*Potamogeton* spp.) were found in 102 stomachs, widgeon grass (*Ruppia maritima*) in 92, eelgrass (*Zostera marina*) in 10, bushy pondweed (*Najas flexilis*) in 9, and horned pondweed (*Zannichellia palustris*) in 8. As in the case of the gadwall, the parts of the pondweeds eaten by the baldpate were almost exclusively leaves and stems, with comparatively few seeds, and birds taken from several different localities evidently had been feeding upon pondweed foliage almost exclusively. One of the plants of this family (*Ruppia maritima*) seems to be well entitled to its common name "widgeon grass," as its foliage is fed upon by the widgeon even more extensively than by the gadwall.

---

[5] Sixty-four of these were examined by W. L. McAtee.

Several "widgeons" shot on the shores of Long Island Slough, southwestern Washington, had eaten considerable quantities of the leaves and rootstocks of eelgrass (*Zostera marina*), a few of the stomachs containing no other food.

GRASSES (GRAMINEAE), 13.9 PER CENT.

The principal grasses taken were switchgrass (*Panicum* spp.), by 11 widgeons, wild rice (*Zizania palustris*), by 5, and saltgrass (*Distichlis spicata*), by 5; rangegrass (*Panicum obtusum*), a little barley (*Hordeum pusillum*), and cultivated rice (*Oryza sativa*) were eaten by one each; and in 16 stomachs were grasses which were not identified. Six full stomachs collected in south central Louisiana in March contained practically nothing besides the remains of tender young shoots of switchgrass. Several from other localities were filled with grass fibers and root stocks, and some contained grass seeds. One from Oregon held more than 1,200 seeds of switchgrass in addition to about 2,800 seeds of another grass which was not determined. The only cultivated grain found was a small quantity of rice taken from a stomach collected in Louisiana in January, when the grain could hardly have been anything but waste. The widgeon has been accused of doing considerable damage to fields of growing grain in spring, but such complaints are not borne out by the present investigation. It is very probable that flocks of the ducks do some little harm in this way, but such depredations are the exception rather than the rule.

ALGAE, 7.71 PER CENT.

Algae, consisting chiefly of musk grass, were found in the stomachs of 25 baldpates. More than two-thirds of this food was taken during the months of April and September, by ducks shot in Wisconsin, Michigan, and Minnesota, probably in migration.

SEDGES (CYPERACEAE), 7.41 PER CENT.

The sedges do not play so important a part in the food of the baldpate as with the gadwall and several other ducks, probably because the seeds are the only parts usually eaten, and this duck evidently cares little for seeds. The sedges eaten by the baldpate were: Three-square (*Scirpus americanus*), by 37; prairie bulrush (*S. paludosus*), 12; river bulrush (*S. fluviatilis*), 5; unidentified bulrushes (*Scirpus* spp.), 24; spike rush (*Eleocharis* sp.), 19; chufa (*Cyperus* sp.), 5; saw grass (*Cladium effusum* and *C. mariscoides*), 6; sedges of the genus *Carex*, 12; *Fimbristylis*, 4; and unidentified sedges, by 20. One duck shot in Chihuahua, Mexico, had swallowed no less than 64,000 seeds of spike rush. As a rule the baldpate does not take sedge seeds freely where pondweeds and other aquatic plants with tender foliage are available.

FROGBIT FAMILY (HYDROCHARITACEAE), 5.75 PER CENT.

The plants of the frogbit family eaten by the baldpate consisted of wild celery (*Vallisneria spiralis*), which was found in 12 stomachs, and waterweed (P*hilotria canadensis*), in 1. Wild celery is a favorite food of the canvas-back, redhead, and other deep-water ducks, but as a rule it is not often found in the stomachs of ducks which do not dive. However, the stomachs of baldpates from several different localities were filled with wild celery leaves. This is very probably due to a peculiar habit which the baldpate has of following the diving ducks and feeding upon the leaves which they bring to the surface. This habit was noted by Wilson and Bonaparte[6] as early as 1831, and has been widely quoted by various writers since then. According to these early ornithologists, "The widgeon is the constant attendant of the celebrated canvass-back duck, so abundant in various parts of the Chesapeake Bay, by the aid of whose labour he has ingenuity enough to contrive to make a good subsistence. The widgeon is extremely fond of the tender roots of that particular species of aquatic plant on which the canvass-back feeds, and for which that duck is in the constant habit of diving. The widgeon, who never dives, watches the moment of the canvass-back's rising, and, before he has his eyes well opened, snatches the delicious morsel from his mouth and makes off." It is probable that these observations are not entirely accurate, as the canvas-back is known to feed chiefly upon the rootstocks of the plant; the baldpate merely avails itself of the leaves thus cut off, brought to the surface, and discarded by the canvas-back.

WATER MILFOILS (HALORAGIDACEAE), 3.48 PER CENT.

Water milfoil (*Myriophyllum* sp.) had been eaten by 24 of the baldpates, and bottle brush (*Hippuris vulgaris*) by 18. Many species of ducks feed upon the seeds of these plants in small numbers, but the baldpate so far as known is the only duck which shows any particular fondness for their foliage. Several stomachs were found to contain the seeds also, and in a very few instances they predominated over the foliage, but the bulk of the food derived from this family of plants consisted of the tender leaves and stems. A series of baldpates from Klamath Falls, Oreg., especially, had partaken of the foliage of *Myriophyllum* in considerable quantities.

DUCKWEEDS (LEMNACEAE), 2.2 PER CENT.

Like the gadwall, the baldpate shows less partiality toward the duckweeds than do some other ducks. The stomachs of three individuals, one each from Wisconsin, Utah, and Oregon, were

---

[6] Wilson, Alexander, and Charles Lucian Bonaparte, Amer. Ornith., III, p. 198, 1831.

nearly filled with the small individual plants, or thalli, of a duckweed (*Lemna* sp.). These plants are very abundant in many of the localities from which the baldpates were taken, but for some reason other foods seemed to appeal to them more strongly.

SMARTWEEDS (POLYGONACEAE), 1.47 PER CENT.

The seeds of water smartweed (*Polygonum amphibium*) were present in 11 baldpate gizzards, those of dock-leaved smartweed (*P. lapathifolium*) in six. Others identified were knotweed (*P. aviculare*), water pepper (*P. hydropiper*), and lady's-thumb (*P. persicaria*), each in two, and mild water pepper (*P. hydropiperoides*) and black bindweed (*P. convolvulus*), each in one. The fact that the seeds of smartweeds are the only edible parts of these plants probably is the reason that they form so small an item of the baldpate's diet.

ARROW-GRASS FAMILY (JUNCAGINACEAE), 0.36 PER CENT.

The arrow-grass family was represented in two baldpate stomachs from the State of Washington; both were nearly full of the seeds of arrow-grass (*Triglochin maritima*). These plants are quite closely related to the pondweeds, but, unlike the pondweeds, their seeds are the only parts eaten by birds.

WATERLILY FAMILY (NYMPHAEACEAE), 0.26 PER CENT; AND HORNWORT FAMILY (CERATOPHYLLACEAE), 0.24 PER CENT.

One stomach from Oregon was nearly filled with 50 of the large seeds of spatterdock (*Nymphaea* sp.) and fragments of many more. Two others contained seeds of watershield (*Brasenia schreberi*), and one the seeds of another waterlily (*Castalia* sp.).

As already stated, the baldpate seems to lack the gadwall's taste for the foliage of coontail (*Ceratophyllum demersum*). Only one bird (taken in Oregon in December) had its stomach full of this plant, and two others had taken a few of the seeds.

MISCELLANEOUS VEGETABLE FOOD, 7.63 PER CENT.

The stomach of one baldpate from lower Chesapeake Bay contained the remains of about 400 seeds of beggar-ticks, or "pitchforks" (*Bidens* sp.). In another from Texas were over 500 seeds of a wild heliotrope (*Heliotropium indicum*), which are often taken by ducks in much smaller numbers; in this instance they furnished 80 per cent of the contents. A stomach from Virginia was filled with the remains of a great many small tubers of arrowhead (*Sagittaria* sp.); one from Massachusetts contained quantities of the leaves of pipewort (*Eriocaulon* sp.); and one from Utah was from a duck which had made a meal of the foliage and seeds of picklegrass (*Salicornia ambigua*). Among other items found in small quantities were bits of the scales from cones of cypress (*Taxodium distichum*), seeds of bur reed (*Sparganium* sp.), myrtle (*Myrica* sp.), saltbush (*Atriplex* sp.), purslane (*Portulaca* sp.), crowfoots (*Ranunculus* spp.), brambles (*Rubus* spp.), clovers (*Melilotus* sp. and *Medicago denticulata*), spurge (*Croton* sp.), sumac (*Rhus* sp.), holly (*Ilex* sp.), water hemlock (*Cicuta* sp.), and many others.

ANIMAL FOOD.

Animal food amounted to 6.77 per cent of the contents of the 229 baldpate stomachs included in the computation. Even this figure is probably unduly large, because the greater part of the animal matter consisted of snails found in the gizzards of a series of ducks from southern Oregon, the only lot of birds found feeding almost exclusively upon such food. More than nine-tenths of the animal food (6.25 per cent of the total) consisted of mollusks, the remainder being made up of insects (0.42 per cent) and miscellaneous matter (0.1 per cent).

MOLLUSKS (MOLLUSCA), 6.25 PER CENT.

Fragments of small bivalves were found in 6 stomachs, and snails (univalves) in 29. As already stated, the greater part of the mollusks were from a number of Oregon ducks, taken along the shores of the Klamath River. Many of them had gorged themselves upon snails, and these constituted practically 100 per cent of the contents of 13 out of the 17 stomachs in the series, of which 7 contained nothing else. Two other baldpates, one from Lake Michigan near Chicago, and the other from Lake Manitoba, Canada, had fed largely upon mollusks.

INSECTS (INSECTA), 0.42 PER CENT.

Insects which amounted to only 0.42 per cent of the food of baldpates included in our investigation probably are eaten to a greater extent during the summer months, especially by the ducklings. No ducklings of this species were available, but there can be little doubt that, like the young of the gadwall, they feed largely upon the adults and larvae of aquatic insects.

More than two-thirds of the insects eaten by the baldpate (0.29 per cent of the whole) were beetles. These included water scavenger beetles (Hydrophilidae), found in 8 stomachs; predacious diving beetles (Dytiscidae), in 2; leaf chafers (Scarabaeidae), in 2; leaf beetles (Chrysomelidae), in 3; weevils (Rhynchophora), in 2; Dermestidae, in 2; and unidentified fragments of beetles, in 16. One gizzard contained about 85 individuals of a species of rove beetle (Staphylinidae), a small, elongated, soft-bodied insect, which is usually very common about decaying animal matter.

Flies and their larvae and pupae furnished 0.09 per cent of the food. Twelve baldpates had eaten midges (Chironomidae); 9, ephydrid flies (Ephydridae); 3, craneflies (Tipulidae); 1, flies of the family

Muscidae; and 1 contained fly remains which were not identified. The larvae of midges are found in immense numbers in stagnant water in many localities, and are often an important food item for water birds.

The remaining insects, amounting to only 0.04 per cent, consisted of a few caddisfly larvae (Phryganoidea), bugs, chiefly water boatmen (Corixidae), a dragonfly nymph, remains of small crickets (*Nemobius* sp.), a small aquatic caterpillar, a few small ants, and unidentified eggs, larvae, and adults of other forms.

MISCELLANEOUS ANIMAL FOOD, 0.1 PER CENT.

Crustaceans furnished less than 1 per cent of the food of the baldpate. They consisted of sand fleas (Amphipoda), bivalved crustaceans (Ostracoda), and a few unidentified forms. One stomach from St. Paul Island, Alaska, was half full of the remains of sand fleas, and contained nothing else. These, together with bits of hydroids, a few spiders and water mites, and the teeth and scales of small fish, made up the remainder of the animal food.

EUROPEAN WIDGEON.

(*Mareca penelope.*)

The European, or red-headed, widgeon is an Old World species but has been noted occasionally at a number of points on the Atlantic coast of North America, and in the North Central and Lake States. There are also several scattered records of its occurrence on the Pacific coast. In appearance the male European widgeon is similar to the baldpate except that the crown is creamy buff instead of white and the remainder of the head and upper part of the neck are reddish brown, with a black area on the chin and throat.

FOOD HABITS.

Not a great deal is known of its food habits in the United States. Sanford,[7] discussing it, says that, "unlike the American baldpate," it is frequently seen on salt water, feeding almost entirely on the short grass growing on the bottom. However, the baldpate also is known to feed commonly in salt water. Only five stomachs of the European widgeon were available for examination. Two of these were from Back Bay, southeastern Virginia; one contained foliage of widgeon grass (*Ruppia maritima*) and eelgrass (*Zostera marina*); the other, only widgeon grass. The third was from the flats of the Susquehanna River near its mouth in northeastern Maryland and contained rootstocks of pondweeds (*Potamogeton* sp.), bits of stems and a few seeds of dodder (*Cuscuta* sp.), and a few seeds of bur-reed (*Sparganium* sp.). The fourth stomach, from the vicin-

---

[7] Sanford, L. C., L. B. Bishop, and T. S. Van Dyke, The Waterfowl Family, p. 91, 1903.

Bul. 862, U. S. Dept. of Agriculture. PLATE III.

GREEN-WINGED TEAL (NETTION CAROLINENSE).

Male on right; female on left.

ity of Currituck Sound, North Carolina, contained leaves of eelgrass. The fifth, from Ipswich, Mass., contained only seeds of salt-marsh bulrush (*Scirpus robustus*). Thus it will be seen that in all probability the food of the European widgeon does not differ materially from that of its American cousin, the baldpate.

GREEN-WINGED TEAL.

(*Nettion carolinense*).

PLATE III.

The green-winged teal, variously known to sportsmen as green-wing, mud teal, winter teal, or red-headed teal, has a very wide distribution, being found in the breeding season from New York, northern Pennsylvania, Michigan, Nebraska, Colorado, and New Mexico northward to the edge of the Barren Grounds; from near Fort Churchill, Hudson Bay, to Kotzebue Sound; and nearly to Point Barrow, Alaska. The main breeding grounds are in west central Canada from Manitoba to Lake Athabaska, and the bird breeds only rarely in the United States east of the Rocky Mountains. It winters commonly in Mexico and the Bahamas, and rarely in Cuba, Jamaica, and Honduras; occasionally south to Tobago. It is also very common in winter in the Southern States, and many individuals remain throughout the winter as far north as they can find open water. It is one of the early ducks to migrate in spring, usually reaching the latitude of New York City during the first week in April, and arriving at the northern limits of its breeding range by about the first of May.

The adult male green-winged teal can best be distinguished by its dark brown head with a patch of metallic green on each side, including the eye, and extending into a crest at the back of the head. It has also a white crescent in front of the wing and a metallic green speculum or wing patch. This wing patch is not so distinct on the female and young. Any of the teals can be distinguished from most of the other ducks by their small size, the green-wing measuring $12\frac{1}{4}$ to 15 inches in length, the blue-wing $14\frac{1}{2}$ to 16 inches, and the cinnamon teal about 17 inches.

FOOD HABITS.

The green-winged teal feeds largely upon the seeds of pondweeds, bulrushes, and other aquatic plants, although it takes also a smaller proportion of such animal food as insects, small crustaceans, and snails. When much disturbed during the daytime, the flocks feed largely at night. The flesh of the green-wing is very palatable, being considered among the best of American ducks, although it is said soon to become less palatable when the birds have been driven to the seashore and feed upon snails and salt-water crustaceans. On account of the fact that the birds have little suspicion of man and fly in compact flocks, affording opportunities for pot shots, the green-winged teal has been greatly reduced in numbers. It is one of our most desirable game birds and should be carefully guarded against further depletion.

VEGETABLE FOOD.

Of the contents of 653 [8] green-winged teal stomachs examined, more than nine-tenths (90.67 per cent) consisted of vegetable matter. By far the largest item of food contributed by any one family of plants came from the sedges, and this amounted to nearly two-fifths (38.82 per cent) of the total food. Next to the sedges, pondweeds are the favorite food supply, contributing 11.52 per cent, while grasses follow closely with 11, then smartweeds 5.25, algae 4.63, duckweeds 1.9, water milfoils 1.11, arrow-grass 0.91, and bur reed 0.85 per cent. The remaining 14.68 per cent is made up of a great number of smaller items.

SEDGES (CYPERACEAE), 38.82 PER CENT.

The sedges form a very constant item of food for the green-winged teal, being found in some form in 530 of 653 stomachs and forming the sole content of 51. Usually the seeds are taken, but practically all parts of the plants are eaten when young and tender. Seeds of bulrushes (*Scirpus* spp.) form the largest item among the sedges, being found in the greatest number of stomachs and represented by several species. Unidentified bulrush seeds were found in 205 stomachs. The most commonly identified species was three-square (*Scirpus americanus*) from 121 stomachs. Seeds of prairie bulrush (*Scirpus paludosus*) were found in 46 stomachs, those of salt-marsh bulrush (*Scirpus robustus*) in 40, *Scirpus cubensis* in 13, and river bulrush (*Scirpus fluviatilis*) in 5. Other genera of sedges represented were *Fimbristylis*, found in 90 stomachs, *Carex* in 72, *Cyperus* 48, spike rush (*Eleocharis*) 45, beaked rushes (*Rhynchospora*) 5, saw grass (*Cladium*) 91, and unidentified sedge seeds in 44. No fewer than 30,000 seeds of a *Cyperus* were found in one stomach and 25,000 in another, while *Eleocharis* and *Fimbristylis* seeds also occasionally reached as high as 1,000 per stomach.

PONDWEEDS (NAIADACEAE), 11.52 PER CENT.

The pondweed group includes the true pondweeds (*Potamogeton* spp.), ditch or widgeon grass (*Ruppia maritima*), horned pondweed (*Zannichellia palustris*), eelgrass (*Zostera marina*), and bushy pondweed (*Najas* spp.), all of which were found in stomachs of the green-winged teal, and seem to form a very important element of their diet. In most cases the seeds alone are taken, but the ducks

---

[8] Two hundred and sixteen of these were examined by W. L. McAtee.

often eat also the stems, leaves, buds, and tubers of some species of *Potamogeton*, leaves and rootstocks of ditchgrass, and parts of the foliage of bushy pondweed, eelgrass, and horned-pondweed. *Potamogeton* (usually seeds) was found in 250 stomachs. In a few instances the species were identified, the most common being sago pondweed (*Potamogeton pectinatus*); but usually it was useless to attempt to identify species by the seeds, as they are so much alike as to be indistinguishable in the worn condition in which they are found in the stomachs. Seeds of this genus, however, even when present in small fragments, are easily distinguished from other seeds by the peculiar curved shape of the cavity which contains the embryo. The seeds of widgeon grass were found in 108 gizzards, and fragments of the leaves were identified from three. Seeds of eelgrass were present in 3 stomachs, bushy pondweed in 27, and horned pondweed in 10. One of the latter stomachs contained more than 1,300 seeds.

GRASSES (GRAMINEAE), 11 PER CENT.

Eighteen species of grass seeds were identified from the birds examined, and unidentified grass seeds were taken from 19 stomachs. Those of the genus *Panicum* were most commonly eaten, being found in 59 gizzards, often constituting a large proportion of the contents, and reaching as high as two or three thousand in number. Another favorite seed was that of barnyard grass, or wild millet (*Echinochloa crus-galli*), which was found in 14 stomachs, and usually formed the bulk of the food whenever it occurred. One duck taken in Louisiana in January had eaten 6,000 seeds of jungle rice (*Echinochloa colona*), both the stomach and gullet being crammed full. Other grass seeds eaten by this teal were wild rice (*Zizania palustris*), taken by 18 birds; cut-grass (*Zizaniopsis miliacea*), by 8; foxtail grasses (*Chaetochloa glauca* and other species), 9; and *Monanthochloë littoralis*, 16. A few kernels of corn had been taken by one bird, and rice by 21. However, all these ducks were collected during the winter months, and the rice and corn were undoubtedly waste grain.

SMARTWEEDS (POLYGONACEAE), 5.25 PER CENT.

Next in order of importance in the food of the green-winged teal come the smartweeds, which form one of the principal items of food of a great many birds. Thirteen species of smartweed were identified, the most important being water smartweed (*Polygonum amphibium*), found in 35 stomachs; dock-leaved smartweed (*P. lapathifolium*), in 29; Opelousas smartweed (*P. opelousanum*), 14; water pepper (*P. hydropiper*), 12; and mild water pepper (*P. hydropiperoides*), 10. The other smartweeds were found in only a few stomachs each, and those taken from 22 other birds were not identified. One duck had eaten 1,630 seeds of knotgrass (*Polygonum aviculare*). Seeds of dock (*Rumex* spp.), another plant of the smartweed family, had been taken by 5 birds.

ALGAE, 4.63 PER CENT.

Musk grass (*Chara* sp.) forms the bulk of the algae taken by the green-winged teal, being found in 89 of the 96 stomachs which contained algae. All parts of the plant are eaten freely, but the ducks seem to be especially fond of the oögonia, very small spherical or egg-shaped objects which form part of the reproductive apparatus and are attached to the whorled leaves. They are usually coated with lime and are rather hard, and consequently often remain in the stomach after all other parts of the plant are digested. Stomachs were found containing thousands of them, and occasionally they constituted the total contents. Musk grasses, of which there are many species, have a very wide distribution, and have been found in duck stomachs from practically all parts of North America.

DUCKWEEDS (LEMNACEAE), 1.9 PER CENT.

The duckweeds, the simplest and smallest of flowering plants, form a rather important element in the food of nearly all ducks which live on plant matter. These plants, at least in the typical genera, consist of merely a frond or leaf floating freely upon the water, with one or more small roots dangling below. The fronds are fleshy and tender, and are scooped up greedily by the ducks. They had been taken by 44 of the 653 green-winged teal examined, and averaged 1.9 per cent of the total food.

WATER MILFOIL FAMILY (HALORAGIDACEAE), 1.11 PER CENT.

The water milfoil family is represented in North America by three genera: Water milfoil (*Myriophyllum*), mermaid-weed (*Proserpinaca*), and bottle brush (*Hippuris*). The seeds of all three of these were present in the series of gizzards examined. Water milfoil seeds had been eaten by 58 birds, those of bottle brush by 16, and those of mermaid-weed by only one.

ANIMAL FOOD.

Insects formed 4.57 per cent of the total food of the green-winged teal, the remainder of the animal food consisting of mollusks, 3.59 per cent; crustaceans, 0.92; and miscellaneous, 0.25; the total amounting to 9.33 per cent.

INSECTS (INSECTA), 4.57 PER CENT.

The largest item of insect food eaten by these ducks was flies (Diptera), which constituted 2.07 per cent of the total. Nearly all of these were in the form of larvae or pupae, the adult flies seldom being caught. Probably those found had been taken from the surface of the water, as it does not seem likely that a duck would be adept at fly-catching. The larvae of midges (Chironomidae) were found in 61 stomachs, sometimes in very large numbers, and formed the bulk of the dipterous food taken. They are abundant in shallow, standing water and slow streams almost everywhere, feeding upon decayed vegetable matter, and evidently are eagerly sought by the ducks. The larvae and pupae of craneflies (Tipulidae), soldierflies (Stratiomyidae), and Ephydridae were also commonly taken.

Although beetles (Coleoptera) formed only 0.65 per cent of the total food, they were represented by a larger number of families and genera than the flies. Those most commonly taken were predacious diving beetles (Dytiscidae), water scavenger beetles (Hydrophilidae), crawling water beetles (Haliplidae), snout beetles and other weevils (Rhynchophora), and ground beetles (Carabidae).

Next in order of importance in the insect food of this teal come the bugs (Hemiptera), with 0.54 per cent, including both the true bugs (Heteroptera) and the cicadas, leafhoppers, etc. (Homoptera). Of the true bugs, water boatmen (Corixidae) were found in 32 stomachs, sometimes in very large numbers; back swimmers (Notonectidae) in 4 stomachs; water striders (Gerridae) in 4; and unidentified bugs in 6. The Homoptera were represented by a single jassid, or leafhopper.

Caddisflies (Phryganoidea) furnished 0.31 per cent of the total food of the birds examined. These were taken in the form of the larvae, or caddis worms, which abound in creeks and ponds, or anywhere in shallow water containing the vegetation upon which the fly larvae feed. They live within silk cases or hollow cylinders made by themselves and covered with a variety of materials, such as grains of sand, bits of leaves or rushes, or pieces of mollusk shell. These cases are open at each end, and the larva pulls itself along by means of three pairs of legs which, with the head, can be protruded from one end. Caddis larvae or cases were found in 46 stomachs, never in very large numbers.

The remaining insect food (1 per cent) was made up of damselflies (Zygoptera), dragonflies (Anisoptera), stoneflies (Plecoptera), bird lice (Mallophaga), grasshoppers (Orthoptera), ant-lions (Neuroptera), moths and butterflies (Lepidoptera), ants, bees, and wasps (Hymenoptera), and a number of miscellaneous unidentified insects and their eggs, pupae, and larvae. Probably the largest single item among these miscellaneous orders of insects was the nymphs of damselflies and dragonflies, identified from 23 stomachs.

MOLLUSKS (MOLLUSCA), 3.59 PER CENT.

Next to insects, mollusks furnished the largest item of animal food for this teal, 3.59 per cent of the total. They were usually found broken, although whole snails were sometimes present. Empty shells or bits of shell probably are often taken by ducks in lieu of gravel to help grind the food, but there is no doubt that the mollusks themselves, and especially snails, are relished by the birds and form an important element in their food. Three genera of snails were identified: *Physa, Neritina,* and *Planorbis*. Unidentified snails were taken from 44 stomachs, and bivalves from only 3. Broken mollusk shells, unclassified, were found in 90 gizzards.

CRUSTACEANS (CRUSTACEA), 0.92 PER CENT.

Small crustaceans, which are abundant in numbers and variety in nearly all streams and bodies of water, whether salt or fresh, are sought by nearly all ducks. They furnished 0.92 per cent of the total food of the green-winged teal, or approximately one-tenth of the animal food. Chief among these were the ostracods, small bivalved crustaceans which might easily be mistaken for minute mollusks. Small shrimplike crustaceans known as amphipods were taken in some numbers, and in one stomach the claws of an unidentified crab were found.

MISCELLANEOUS ANIMAL FOOD, 0.25 PER CENT.

A few spiders and mites (class Arachnida), centipeds (Myriapoda), fish scales, minute aquatic animalculae, and other insignificant items form the remainder of the green-winged teal's animal food.

BLUE-WINGED TEAL.

(Querquedula discors.)

PLATE IV.

The blue-winged teal, blue-wing, or summer teal is slightly more restricted in its distribution than the green-wing. Although it has been recorded as breeding in Rhode Island, Maine, New Brunswick, Nova Scotia, Newfoundland, Quebec, Ontario, and New York, and as far south as northern Ohio, southern Indiana, southern Colorado, New Mexico, Texas, Utah, northern Nevada, and central Oregon, it is not common east of the Alleghany Mountains nor on the Pacific slope. Its principal summer home is in the interior of North America between the Rocky Mountains and the Great Lakes, from northern Illinois and Nebraska north to Saskatchewan. Its principal range extends north to British Columbia, and it occurs also rarely north to Alaska, Alberta, and about Great Slave Lake. In winter, blue-winged teals are found throughout northern South America south to Brazil, Ecuador, Peru, and Chile; they occur abundantly in Central America, Mexico, and the West Indies; and in the United States they are found near the Gulf, and as far north as North Carolina, and (sparingly) southern Indiana and southern Illinois. Unlike the green-winged teal, this is one of the least hardy of our ducks, migrating late in spring

BLUE-WINGED TEAL (QUERQUEDULA DISCORS).

Male on left; female on right.

and early in fall. It usually arrives in central Iowa during the last week in March, and at Aweme, Manitoba, about a month later. In the fall migration it reappears throughout the northern half of the United States during the month of August and reaches the Gulf of Mexico about the middle of September. In habits it is very similar to the green-winged teal, and like that bird its numbers have been greatly diminished in recent years on account of its slight fear of man and the consequent ease with which it may be shot by even inexperienced sportsmen. It is especially rare in most of the States east of the Alleghenies, and great care should be taken in some localities to see that it is not entirely wiped out.

In general appearance the blue-winged teal is similar to the greenwing, having also a green speculum, which, however, is supplemented by a light-blue shoulder patch, separated from the green by a narrow white line. The adult male also lacks the white mark before the wing, which is present in the green-winged teal, but has a large white crescent on each side of the face in front of the eye.

FOOD HABITS.

To determine the food habits of the blue-winged teal, 319 [9] stomachs were examined, collected from 29 States and 4 Canadian Provinces during a period of 31 years, and in every month but January. As might be expected, the greatest numbers were collected in the fall, during the months of September, October, and November, making the average percentages of various kinds of foods for those months more accurate than for the remainder of the year. Rather large series were collected in Wisconsin (58), Florida (46), Maine (40), and North Dakota (36); the remaining stomachs were fairly evenly distributed. The character of the contents of the stomachs from the States furnishing the largest numbers was not such as to influence unduly the final averages.

VEGETABLE FOOD.

About seven-tenths (70.53 per cent) of the blue-winged teal's food consists of vegetable matter. Of this about three-fourths is included in four families of plants. Sedges (Cyperaceae), with 18.79 per cent; pondweeds (Naiadaceae), 12.6; grasses (Gramineae), 12.26; and the smartweeds (Polygonaceae), 8.22. The remainder of the plant food is made up of algae, 2.95 per cent; waterlilies (Nymphaeaceae), 1.37; rice and corn, 0.98; water milfoils (Haloragidaceae), 0.71; bur reeds (Sparganiaceae), 0.38; madder family (Rubiaceae), 0.35; and miscellaneous, 11.92 per cent.

---

[9] Ninety of these were examined by W. L. McAtee.

SEDGES (CYPERACEAE), 18.79 PER CENT.

The sedges are grasslike or rushlike plants which grow in marshes or on the borders of ponds and streams. Ducks are especially fond of their seeds, which usually are small and hard and have a starchy interior. The family of sedges is a very large one, comprising about 3,200 species, widely distributed. The seeds most often found in duck stomachs are those of the bulrushes (*Scirpus* spp.), and the case of the blue-winged teal is no exception to this rule. Unidentified bulrush seeds were found in 184 stomachs, those of river bulrush (*S. fluviatilis*) in 18, three-square (*S. americanus*) in 10, prairie bulrush (*S. paludosus*) in 7, and great bulrush (*S. validus*) and salt-marsh bulrush (*S. robustus*) in 2 each. Other sedges taken were those of the genus *Carex*, found in 59 stomachs; saw grass (*Cladium effusum* and *C. mariscoides*), in 55; chufa (*Cyperus* spp.), in 45; spike rush (*Eleocharis* spp.), in 33; beaked rush (*Rhynchospora* sp.), in 2; and sedges of the genera *Fimbristylis*, in 40; and *Dulichium*, in 2. Unidentified sedge seeds or bits of the plants were taken by 27 birds.

PONDWEEDS (NAIADACEAE) 12.6 PER CENT.

In 33 of the stomachs examined the seeds or other parts of pondweeds formed from 95 to 100 per cent of the total food contents. The true pondweeds (*Potamogeton* spp.) had been taken by 151 birds, widgeon grass (*Ruppia maritima*) by 87, bushy pondweed (*Najas flexilis* and *N. marina*) by 18, eelgrass (*Zostera marina*) by 3, and horned pondweed (*Zannichellia palustris*) by 2. One stomach held over 700 of the hard black seeds of widgeon grass. Most ducks feed upon all parts of pondweed plants, and the blue-winged teal seems to pay much attention to the leaves and stems as well as the seeds.

GRASSES (GRAMINEAE), 12.26 PER CENT.

Of the 319 blue-winged teals examined, only 13 had eaten cultivated grain. One of these, obtained in Kansas in April, had its gizzard filled with 19 kernels of corn and fragments of more, but corn taken at that time of year could hardly have been anything but waste. The other 12 birds had eaten rice, and as all were collected in Florida in November, this, too, was undoubtedly waste grain. Of the wild grasses the favorites were wild rice (*Zizania palustris*), taken by 22 birds; switchgrass (*Panicum* sp.), by 18; the foxtails (*Chaetochloa glauca, C. viridis*, and others), by 14; rice cut-grass (*Homalocenchrus oryzoides*), by 9; and *Monanthochloë littoralis*, by 13. Other species less often taken were meadow grass (*Puccinellia nuttalliana*), barnyard grass (*Echinochloa crus-galli*), cut-grass (*Zizaniopsis miliacea*), rushgrass (*Sporobolus* sp.), and salt-marsh grass (*Spartina* sp.).

FOOD HABITS OF SHOAL-WATER DUCKS.

SMARTWEEDS (POLYGONACEAE), 8.22 PER CENT.

Two of the blue-winged teals had eaten seeds of dock (*Rumex* sp.). All other seeds of this family taken were of the true smartweeds (*Polygonum* spp.). These were represented by 9 species, and 16 stomachs contained unidentified smartweed seeds. Mild water pepper (*Polygonum hydropiperoides*), which was found in 31 stomachs; water smartweed (*P. amphibium*), in 27; and dock-leaved smartweed (*P. lapathifolium*), in 26, were the kinds most often found. Other species taken were prickly smartweed (*P. sagittatum*), lady's-thumb (*P. persicaria*), water pepper (*P. hydropiper*), Opelousas smartweed (*P. opelousanum*), Pennsylvania smartweed (*P. pennsylvanicum*), and dense-flowered smartweed (*P. portoricense*).

ALGAE, 2.95 PER CENT.

The greater part of the seaweeds taken consisted of musk grass (*Chara* spp.). Several stomachs collected in Wisconsin, North Dakota, and Florida were nearly full of this alga, chiefly the oögonia, or reproductive cells. Altogether, musk grass was found in 31 stomachs, and unidentified marine algae, or seaweeds, in 4.

WATERLILIES (NYMPHAEACEAE), 1.37 PER CENT.

Waterlily seeds had been taken by 27 of these teals. Fourteen had eaten seeds of white waterlilies (*Castalia* sp.), and the other 13 had eaten those of the small purple waterlily known as water shield (*Brasenia schreberi*). Most of the white waterlily seeds were found in the stomachs of a series of ducks collected in Florida. One of these, together with the bird's gullet, which was also full, contained 1,600 seeds and fragments of many more.

WATER MILFOILS (HALORAGIDACEAE), 0.71 PER CENT.

The plants of the family Haloragidaceae have a very wide geographic distribution. They are chiefly aquatic, and have hard, nutlike seeds which persist for some time in bird stomachs. The three North American genera were represented in the stomachs examined, bottle brush (*Hippuris vulgaris*) in 8, mermaid weed (*Proserpinaca* sp.) in 5, and water milfoil (*Myriophyllum* sp.) in 44.

BUR REEDS (SPARGANIACEAE), 0.38 PER CENT.

The seeds of bur reed (*Sparganium* sp.) had been eaten by 39 of the blue-winged teals examined, but usually were found in small numbers.

MADDER FAMILY (RUBIACEAE), 0.35 PER CENT.

The madder family was rather sparingly represented by seeds of buttonbush (*Cephalanthus occidentalis*), found in 5 stomachs; bedstraw, or cleavers (*Galium* sp.), in 10; and rough buttonweed (*Diodia teres*) in 1.

MISCELLANEOUS VEGETABLE FOOD, 11.95 PER CENT.

A large number of minor items of vegetable food were classified as miscellaneous. Probably the largest of these consisted of plants of the duckweed family (Lemnaceae). Although found in only 14 stomachs, they constituted nearly 100 per cent of the contents of several. Each of three stomachs collected in Iowa in August contained more than a thousand of the small plants of a duckweed (*Lemna* sp.). Twenty-eight other families of plants were represented, the most important being the aster family (Compositae), the water plantain family (Alismaceae), the parsley family (Umbelliferae), crowfoot family (Ranunculaceae), borage family (Boraginaceae), myrtle family (Myricaceae), rose family (Rosaceae), hornwort family (Ceratophyllaceae), and the vervain family (Verbenaceae).

ANIMAL FOOD.

Animal matter constitutes 29.47 per cent of the total food of the blue-winged teal, which is more than three times the percentage of animal food eaten by the green-wing. Over half of this (16.82 per cent) is mollusks, the remainder being made up of insects, 10.41 per cent; crustaceans, 1.93, and miscellaneous, 0.31 per cent.

MOLLUSKS (MOLLUSCA), 16.82 PER CENT.

The greater part of the shellfish found in the stomachs examined probably consisted of snails, although small bivalves also had been commonly taken, and in a majority of cases the shells had been so thoroughly crushed by the powerful gizzards of the ducks as to make it impracticable to distinguish between the fragments of bivalves and univalves. However, 15 species of the latter were identified, and 2 of the former. Unidentified univalve shells were found in 31 stomachs and unidentified bivalves in 2, while fragments of mollusk shells taken from 106 stomachs were not classified. The full stomach of a duck collected in an Iowa swamp in August, 1907, contained thousands of snail eggs, amounting to 54 per cent of its contents.

INSECTS (INSECTA), 10.41 PER CENT.

The items of insect food of the blue-winged teal, in the order of their importance, are caddis larvae (together with their cases), beetles and their larvae, dragonflies and damselflies (chiefly in the nymph stage), bugs, flies (chiefly larvae), and a small percentage of miscellaneous insects.

The larvae of caddisflies (Phryganoidea) or their cases were found in 37 stomachs, and amounted to 4.5 per cent of the total food. The greater part of these were found in a series of stomachs collected in Florida, some of which were over half filled with the fragments of caddis cases.

Beetles (Coleoptera) amounted to 2.62 per cent of the food of the blue-winged teal, or less than one-tenth of the total animal matter eaten. Ten species of predacious diving beetles (Dytiscidae) were noted, 7 of ground beetles (Carabidae), 5 of water scavenger beetles (Hydrophilidae), 4 of crawling water beetles (Haliplidae), 3 of leaf chafers (Scarabaeidae), 3 of leaf beetles (Chrysomelidae), 2 each of snout beetles (Curculionidae) and billbugs (Calandrinae), and 1 each of whirligig beetles (Gyrinidae), shining carrion beetles (Histeridae), pill beetles (Byrrhidae), and mud beetles (Heteroceridae); while many individuals of most of these families were found which, on account of their fragmentary condition, could not be further identified. Unclassified beetle remains were found in 50 stomachs.

The nymphs or young of damselflies (Zygoptera) and dragonflies (Anisoptera) live in the water and afford delicate morsels for ducks. Twenty-two of the blue-winged teals had eaten nymphs of dragonflies and two those of damselflies, while three stomachs contained remains of nymphs which were not identified.

Bugs (Heteroptera and Homoptera) constituted 0.86 per cent of the birds' diet. These represented 10 families, besides the remains of a few bugs which were not identified. Water boatmen (Corixidae) had been eaten by 43 birds, creeping water bugs (Naucoridae) by 15, back swimmers (Notonectidae) by 12, water striders (Gerridae) and broad-shouldered water striders (Veliidae) by 2 each, and negro bugs (Corimelaenidae), stink bugs (Pentatomidae), giant water bugs (Belostomatidae), planthoppers (Fulgoridae), and leafhoppers (Jassidae) by 1 each.

Only 0.65 per cent of the blue-winged teal's food consisted of two-winged flies and their larvae and pupae. Six families were represented, and unidentified larvae or pupae were taken from 8 stomachs. The larvae of soldierflies (Stratiomyidae) and midges (Chironomidae) were present in 11 gizzards each, while those of flower flies (Syrphidae) had been eaten by 4 birds, and Anthomyiidae, Ephydridae, and black flies (Simuliidae) by 1 each.

The miscellaneous insect food consisted of unidentified fragments of insects, a grasshopper or two, 3 small moth cocoons, a few ants, insect eggs, etc.

CRUSTACEANS (CRUSTACEA), 1.93 PER CENT.

Crustaceans furnished 1.93 per cent of the contents of all the blue-winged teal gizzards examined, and consisted of beach fleas, scuds, etc. (Amphipoda), found in 7 stomachs; small bivalved crustaceans (Ostracoda), in 8; and stalk-eyed crustaceans (Decapoda), in 2. The last-mentioned order includes the claw of a crab found in one stomach and a sand shrimp (*Crangonyx gracilis*) in the other. Two North Carolina stomachs collected in March were nearly filled with beach fleas, or amphipods. Crustaceans which had been taken by 5 other teals were too fragmentary for identification.

MISCELLANEOUS ANIMAL FOOD, 0.31 PER CENT.

The miscellaneous animal food, which amounted to only 0.31 per cent, consisted principally of the remains of a few minnows and other small fishes, a few spiders, and several tiny water mites, or hydrachnids.

CINNAMON TEAL.

Querquedula cyanoptera.

PLATE V.

The cinnamon teal is a western bird, its breeding range in North America extending from eastern Wyoming and western Kansas west to the Pacific coast, and from southern British Columbia and southwestern Alberta south to northern Lower California, northern Mexico, southern New Mexico, and central western Texas. Its distribution is very remarkable in that it not only breeds in the Northern Hemisphere, but also over a large area in South America, the two colonies being separated by a zone about 2,000 miles wide in which the species is practically unknown. The cinnamon teal of North America migrates in winter only a short distance south of its breeding range in Mexico and is found at this season as far north as Brownsville, Tex., central New Mexico, southern Arizona, and Tulare Lake, California. The South American birds migrate slightly northward after nesting, but the breeding seasons of the two colonies are, of course, reversed.

The male cinnamon teal differs from the blue-wing in appearance in having a blackish area on the top of the head, and chestnut or cinnamon brown on the remainder of the head, neck, and underparts, giving it the local name of red-breasted teal.

FOOD HABITS.

Only 41 stomachs of the cinnamon teal were available for examination. These were collected during the eight months from March to October, and from the States of Colorado, Utah, Arizona, Montana, Oregon, and California, the bulk being from Utah and California. Although the number is too small to furnish an accurate estimate of the percentages of various foods taken, nevertheless the results are of value in showing that this species probably does not differ materially in habits from the other two North American teals.

VEGETABLE FOOD.

Like the green-wing and the blue-wing, the cinnamon teal lives mainly upon vegetable food, this comprising about four-fifths (79.86 per cent) of the total contents of the stomachs examined. And like the other teals its two principal and most constant items of food are the

Cinnamon Teal (Querquedula cyanoptera).
Male on left; female on right.

seeds and other parts of sedges (Cyperaceae) and pondweeds (Naiadaceae). These two families of plants furnished 34.27 and 27.12 per cent, respectively, of the bird's entire diet. The grasses (Gramineae) amounted to 7.75 per cent; smartweeds (Polygonaceae), to 3.22; mallows (Malvaceae), 1.87; goosefoot family (Chenopodiaceae), 0.75; water milfoils (Haloragidaceae), 0.37; and miscellaneous, 4.51.

SEDGES (CYPERACEAE), 34.27 PER CENT.

Twelve birds had eaten seeds of prairie bulrush (*Scirpus paludosus*), 3 those of three-square (*S. americanus*), and the stomachs of 17 contained seeds of unidentified bulrushes. Seeds of spike rush (*Eleocharis* sp.) had been taken by 10, seeds of *Carex* by 5, and unidentified sedges by 7.

PONDWEEDS (NAIADACEAE), 27.12 PER CENT.

The pondweeds eaten consisted of seeds of true pondweeds (*Potamogeton* spp.), found in 33 stomachs; widgeon grass (*Ruppia maritima*), in 16; and horned pondweed (*Zannichellia palustris*), in 10. One duck had eaten over 400 large seeds of *Potamogeton*, and another 950 seeds of widgeon grass.

GRASSES (GRAMINEAE), 7.75 PER CENT.

The seeds of *Monanthochloë littoralis* were identified from 5 stomachs. Other grass seeds and bits of grass fiber were found in 3.

SMARTWEEDS (POLYGONACEAE), 3.22 PER CENT.

Seeds of smartweed (*Polygonum lapathifolium*) had been eaten by 3 of the cinnamon teals, those of lady's-thumb (*P. persicaria*) by 1, and unidentified smartweeds by 3. Two birds had taken seeds of dock (*Rumex* sp.).

MALLOW FAMILY (MALVACEAE); GOOSEFOOT FAMILY (CHENOPODIACEAE); AND WATER MILFOIL FAMILY (HALORAGIDACEAE), 2.99 PER CENT.

Two stomachs contained unidentified seeds of the mallow family, amounting to 1.87 per cent of the whole. Another contained fragments of several hundred seeds of a pigweed (*Chenopodium* sp.), furnishing 0.75 per cent. Three birds had eaten seeds of bottle brush (*Hippuris vulgaris*) and 2 those of water milfoil (*Myriophyllum* sp.), together amounting to 0.37 per cent of the total.

MISCELLANEOUS VEGETABLE FOOD, 4.51 PER CENT.

A few seeds each of bur reed (*Sparganium* sp.), amaranth (*Amaranthus* sp.), yellow water-crowfoot (*Ranunculus delphinifolius*), bur clover (*Medicago denticulata*) and other clovers (*Medicago* sp. and *Trifolium* sp.), California sumach (*Rhus laurina*), heliotrope (*Heliotropium indicum*), and cleavers (*Galium* sp.), and traces of musk grass (*Chara* sp.), made up the remainder of the bird's vegetable food.

ANIMAL FOOD.

The 41 cinnamon teals examined had made of animal matter 20.14 per cent of their food. This consisted of insects, 10.19 per cent; mollusks, 8.69 per cent; and a few small miscellaneous items, 1.26 per cent.

INSECTS (INSECTA), 10.19 PER CENT.

Over half the insect food of the series of cinnamon teals (5.4 per cent of the whole) consisted of beetles (Coleoptera). Disregarding several unidentified fragments, only four families were represented, the predacious diving beetles (Dytiscidae), water scavenger beetles (Hydrophilidae), leaf beetles (Chrysomelidae), and snout beetles (Curculionidae).

The bugs (Heteroptera) amounted to 2.97 per cent, and consisted entirely of water boatmen (Corixidae). These are small brown or gray mottled bugs, with oarlike legs well fitted for swimming; they frequent the lakes, ponds, and streams throughout the greater part of North America, and are commonly eaten by many species of water birds. As they are very good swimmers, it must require quick work on the part of the ducks to catch them. They were found in 11 of the 41 stomachs.

Remains of dragonflies (Anisoptera) were found in two gizzards, and a nymph of a dragonfly or a damselfly in another. The dragonflies and damselflies (Zygoptera) together constitute the superorder Odonata, which furnished 0.92 per cent of the food of the cinnamon teal.

The flies (Diptera) taken were mostly larvae, and amounted to 0.62 per cent. Flies of at least four families—the midges (Chironomidae), soldierflies (Stratiomyidae), flower flies (Syrphidae), and brine flies (Ephydridae)—were included. A few insect eggs, bits of the cases of caddis larvae (Phryganoidea), two small hymenopterous cocoons, and the remains of an ant, together amounting to 0.28 per cent, made up the remainder of the insect food.

MOLLUSKS (MOLLUSCA), 8.69 PER CENT.

Four of the cinnamon teals had fed upon snails and two upon small bivalves, and the stomachs of 15 contained fine fragments which were not classified. Altogether, mollusks amounted to 8.69 per cent of the bird's diet, a proportion considerably greater than that of the green-winged teal, but only about half as great as that of the blue-wing.

MISCELLANEOUS ANIMAL FOOD, 1.26 PER CENT.

The stomach of a young bird collected near Great Salt Lake, Utah, in July, was half filled with fine feathers. These, together with a few water mites (Hydrachnidae), bivalved crustaceans (Ostracoda), and a small quantity of unidentified matter from other stomachs, all of which amounted to 1.26 per cent, made up the remainder of the animal food of the species.

PINTAIL.

(Dafila acuta.)

Plate VI.

The pintail breeds abundantly along the northern border of the United States from Lake Superior almost to the Pacific, and northward to the Arctic coast northwest of Hudson Bay and west to Alaska. It is uncommon as a breeder east of a line drawn from the western side of Hudson Bay to the western shore of Lake Michigan, and south of the northern tier of States except on the Great Plains. It breeds also south to northern Illinois, southern Colorado, and southern California, winters as far south as Cuba and Panama, and abundantly in the southern half of the United States. The species breeds also in the northern portions of the Old World and migrates south in winter to northern Africa and southern Asia.

The pintail is easily recognized by its long neck as well as by the long, pointed middle tail feathers from which it derives most of its common names. In addition to "pintail" it is sometimes known locally as "sprig," "sprig-tail," "sharp-tail," and "spike-tail." A cinnamon-brown wing bar, present in both sexes, also is distinctive. The adult male has a very dark brown head, a white stripe on each side of the neck, and the sides and back finely marked with black and white wavy lines.

FOOD HABITS.

In its general habits the pintail quite closely resembles the mallard, although it probably spends less time feeding on dry land remote from the water. It is not particularly adept at diving, but nevertheless obtains much of its food from under the surface and often from the bottom in shallow water, by tipping-up for it. It nests in low meadows or sloughs, frequently some distance from water. The female is very solicitous in the care of her young, attempting to decoy an intruder away from them by playing wounded, or to distract his attention by circling around and quacking loudly. In autumn, pintails usually gather in good-sized flocks either by themselves or with other shallow-water ducks, and are much sought after for their flesh, which is very palatable.

The stomachs of 790 [10] pintails, collected from practically all parts of North America from Alaska and Hudson Bay to California, Texas, and Florida, were available for this investigation. The largest numbers were taken in the States of Louisiana (172), Washington (139), Texas (110), Utah (91), Florida (44), and North Carolina (42), the remainder being well scattered. Of the total number, data on the contents of 769, representing the months from September to March, inclusive, were used in computing averages.

---

[10] Two hundred and thirty-seven of these were examined by W. L. McAtee.

VEGETABLE FOOD.

Vegetable matter constitutes about seven-eighths (87.15 per cent) of the total food of the pintail. This is made up of the following items: Pondweeds, 28.04 per cent; sedges, 21.78; grasses, 9.64; smartweeds and docks, 4.74; arrowgrass, 4.52; musk grass and other algae, 3.44; arrowhead and water plantain, 2.84; goosefoot family, 2.58; waterlily family, 2.57; duckweeds, 0.8; water milfoils, 0.21; and miscellaneous vegetable food, 5.99 per cent.

PONDWEEDS (NAIADACEAE), 28.04 PER CENT.

The pondweed, the family of plants which furnishes the largest item of food for the pintail, is the favorite also of several other species of ducks, including the gadwall and the baldpate. The latter two species, however, partake very largely of the leaves and stems of the plants, while the pintail prefers the seeds. Of the whole number of stomachs of the pintail, 254 contained seeds or other parts of widgeon grass (*Ruppia maritima*), at least four of them with from 1,000 to 1,300 seeds each. Two others contained about 2,800 seeds each of unidentified species of true pondweed (*Potamogeton*), and another held over 2,000 seeds of horned pondweed (*Zannichellia palustris*). Seeds of sago pondweed (*Potamogeton pectinatus*), small pondweed (*P. pusillus*), leafy pondweed (*P. foliosus*), and curly pondweed (*P. diversifolius*) were identified in a few instances, but pondweed seeds of which the species could not be determined were found in 271 stomachs. Seeds of horned pondweed were present in 18 stomachs, those of bushy pondweed (*Najas flexilis*) in 23, while the seeds, and occasionally leaves, of eelgrass (*Zostera marina*) were found in 93, and amounted to 4.03 per cent of the pintail's total food. This item was most abundant in the stomachs of a series of birds from the southwestern coast of Washington.

SEDGES (CYPERACEAE), 21.78 PER CENT.

The seeds of sedges are second only to the pondweeds in importance in the food of the pintail. The plants of this family often are semisubmerged, or grow in marshy situations. Probably most of the seeds are taken from the water after they have ripened and fallen, although no doubt a great many are picked from the shorter plants and those which are bent low over the water. Seeds of the common three-cornered bulrush, or three-square (*Scirpus americanus*), were identified from 155 of the pintail stomachs, those of prairie bulrush (*S. paludosus*) from 84, salt-marsh bulrush (*S. robustus*) from 29, *Scirpus cubensis* from 8, river bulrush (*S. fluviatilis*) from 3, and unidentified bulrushes from 154. Seeds of saw grass (*Cladium effusum*), spike rush (*Eleocharis* sp.), chufa (*Cyperus* sp.), beaked rush (*Rhynchospora* sp.), and sedges of the genera *Fim-*

Bul. 862, U. S. Dept. of Agriculture. PLATE VI.

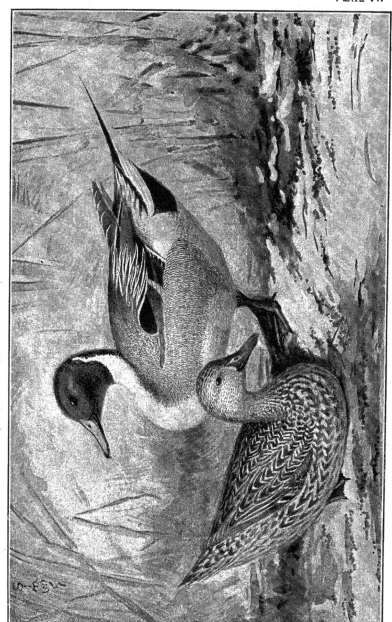

PINTAIL (DAFILA ACUTA).
Male on right; female on left.

bristylis and *Carex* also were taken very frequently by these ducks. As many as 1,500 seeds of beaked rush, 3,000 of the three-square, 3,600 of spike rush, and 9,000 of *Scirpus cubensis* were counted from various single stomachs, while one contained more than 150 of the large, horned seeds of beaked rush (*Rhynchospora corniculata*).

GRASSES (GRAMINEAE), 9.64 PER CENT.

Grass remains in the pintail's food consist also very largely of seeds. Many of the stomachs contained the seeds of switchgrass (*Panicum* sp.), barnyard grass or wild millet (*Echinochloa crus-galli*), wild rice (*Zizania palustris*), cut grass (*Zizaniopsis miliacea*), salt grass (*Distichlis spicata*), and foxtail (*Chaetochloa* sp.), often in very large numbers. One was found to contain 11,500 seeds of barnyard grass, and another held nearly 4,000 of switchgrass. Included in the grass family are cultivated rice (*Oryza sativa*) and other grains. Rice was found in 52 of the pintail stomachs—24 of them from Texas, 18 from Louisiana, and 10 from Florida. Many were crammed with rice kernels, and contained nothing else, but all were taken during the months of November, December, and February, so there can be no doubt that it was all waste grain. Of the other cultivated grains, corn was found in 3 stomachs, wheat in 3, barley in 3, and oats in 1. This also probably was waste grain, with the exception of oats, taken in North Dakota in June, and a few grains of wheat taken in Colorado in March.

SMARTWEEDS AND DOCKS (POLYGONACEAE), 4.74 PER CENT.

Many of the plants of the family Polygonaceae grow in or near water and their seeds are eaten by a great many different kinds of birds. The seeds of mild water pepper (*Polygonum hydropiperoides*), found in 29 stomachs, and of water smartweed (*P. amphibium*), in 23, seem to be the favorites with the pintail. Several also had taken seeds of prickly smartweed (*P. sagittatum*), knotweed (*P. aviculare*), Pennsylvania smartweed (*P. pennsylvanicum*), water pepper (*P. hydropiper*), and other smartweeds (*P. lapathifolium, P. persicaria,* and *P. opelousanum*). About 12,500 seeds of a smartweed, *Polygonum punctatum*, were found in the crammed gullet and gizzard of one pintail from Alabama. Seeds of dock (*Rumex* sp.) were present in 7 gizzards.

ARROW-GRASS (*Triglochin maritima*), 4.52 PER CENT.

Arrow-grass is a plant of the marshes, muddy shores, or low meadows near salt water. The seeds are borne on erect spikes, and are the only parts of the plants commonly eaten by the birds. Eighty-eight pintail stomachs, all from the Pacific coast of Washington, contained these seeds. They were often present in large masses, and together with the seeds of eelgrass formed the bulk of the contents of the stomachs from that region, all of which were taken during the months of October, November, and December.

MUSK GRASS (*Chara*) AND OTHER ALGAE, 3.44 PER CENT.

The stomachs of 49 pintails held at least traces of musk grass, and 8 contained other algae. A few were filled with algae alone. The fact that a duck has been feeding upon musk grass often can be detected by the presence of the small, hard reproductive cells (oögonia) which persist in the stomach after all other parts have been digested.

ARROWHEAD AND WATER PLANTAIN (ALISMACEAE), 2.84 PER CENT.

Many of the pintails shot on the delta of the Mississippi River, Louisiana, had been feeding on the tubers of an arrowhead (*Sagittaria platyphylla*), which is very abundant on the mud flats there. These tubers are known as the "delta potato," and are one of the important duck foods of that region.[11] The seeds of arrowheads occasionally are taken also, as well as the tubers of other species. Seeds of water plantain (*Alisma plantago-aquatica* et al.) were found in 6 stomachs.

GOOSEFOOT FAMILY (CHENOPODIACEAE), 2.58 PER CENT.

Of a series of 35 pintail stomachs collected from a point on the Gulf coast of Texas in October, 33 contained seeds of glasswort (*Salicornia ambigua*), a low, fleshy, leafless plant growing in dense colonies on mud flats too saline for other vegetation. These seeds amounted to 100 per cent of the contents of 25 of the stomachs, and averaged 93 per cent in the series of 35. Two of the stomachs contained no fewer than 28,000 seeds each. Five pintails had eaten seeds of pigweed (*Chenopodium* sp.) and one those of saltbush (*Atriplex* sp.).

WATERLILY FAMILY (NYMPHAEACEAE), 2.57 PER CENT.

The stomachs of 34 pintails contained seeds of water shield (*Brasenia schreberi*), 10 the seeds of waterlilies of the genus *Castalia*, and 6 those of the genus *Nymphaea*, which includes the yellow pondlily. One Florida stomach contained 142 seeds of water shield, 4 of a species of *Castalia*, and the remains of a large number of a species of yellow pondlily.

DUCKWEEDS (LEMNACEAE), 0.8 PER CENT.

Only 15 pintails had eaten duckweeds (*Lemna* sp.). These small plants so relished by the wood duck, form one of the lesser items in the food of the pintail.

---

[11] See Bull. No. 465, U. S. Dept. Agr., pp. 21-24, 1917.

WATER MILFOILS (HALORAGIDACEAE), 0.21 PER CENT.

Water milfoils are usually submerged plants, bearing hard, nutlike seeds in the axils of the finely dissected leaves. The seeds, and sometimes bits of the leaves, are picked off occasionally by ducks. Seeds of water milfoil (*Myriophyllum* sp.) were found in 24 of the pintail stomachs, of bottle brush (*Hippuris vulgaris*) in 12, and mermaid weed (*Proserpinaca* sp.) in 6.

MISCELLANEOUS VEGETABLE FOOD, 5.99 PER CENT.

The seeds of many species of plants not already mentioned were found in stomachs of pintails, sometimes in comparatively large series, but then only in small numbers. In this category are seeds of myrtles or bayberries (*Myrica* spp.), found in 45 gizzards; brambles (*Rubus* spp.), in 43; elders (*Sambucus* spp.), in 32; bur reed (*Sparganium* sp.), in 32; wild heliotrope (*Heliotropium indicum*), in 28; hornwort or coontail (*Ceratophyllum demersum*), in 23; crowfoots (*Ranunculus* spp.), in 17; pigweeds (*Amaranthus* spp.), in 15; hawthorns (*Crataegus* spp.), in 12; grapes (*Vitis* spp.), in 11; mallows (Malvaceae), in 10; and many others. Bits of spruce needles (*Picea* sp.) had been picked up, probably from the water, by 21 of the ducks; four pintails had eaten leaves or rootstocks of wild celery (*Vallisneria spiralis*). The remainder of the miscellaneous vegetable food consisted entirely of seeds.

ANIMAL FOOD.

The animal portion, 12.85 per cent, of the food of the pintail was made up of mollusks, 5.81 per cent; crustaceans, 3.79 per cent; insects, 2.85 per cent; and miscellaneous, 0.4 per cent.

MOLLUSKS (MOLLUSCA), 5.81 PER CENT.

Mollusks were found in 326 of the 790 pintail stomachs examined. In 44 they constituted 50 per cent or more of the contents, and in 5 they reached 100 per cent. Practically all of a large series of stomachs from the Pacific coast of Washington contained mollusk shells. Univalves (Gastropoda) predominated over bivalves (Pelecypoda) in the food of the pintail, but large numbers of small species and young of the latter were taken. In many instances mollusks could not be identified because of the great efficiency of the duck stomachs as grinding machines.

CRUSTACEANS (CRUSTACEA), 3.79 PER CENT.

Remains of crabs were found in 28 of the pintail stomachs, crawfish in 4, and shrimps in 2. Ten ducks shot near the eastern end of Long Island, New York, had been feeding very largely upon a small crab (*Hexapanopeus angustifrons*) which frequents the beaches there. One gizzard contained the claws and other remains of at least 30 of these crabs. When crawfish are eaten by ducks, small planoconvex masses of calcareous matter known as gastroliths, supposed to be used as material for a new exoskeleton after moulting, often are found, with the fingers or claws persisting in the stomach after other parts of the crawfish are digested. Small bivalved crustaceans (Ostracoda) had been taken by 38 of the pintails, sand fleas or beach fleas (Amphipoda) by 20, and sowbugs (Isopoda) by 5.

INSECTS (INSECTA), 2.85 PER CENT.

Beetles (Coleoptera) amounting to 0.93 per cent of the total food of the pintail, consisted largely of three families, the predacious diving beetles (Dytiscidae), water scavenger beetles (Hydrophilidae), and ground beetles (Carabidae). Others represented were the snout beetles (Curculionidae), leaf beetles (Chrysomelidae), leaf chafers (Scarabaeidae), crawling water beetles (Haliplidae), rove beetles (Staphylinidae), click beetles (Elateridae), flat bark beetles (Cucujidae), and tiger beetles (Cicindelidae). Besides these there were the fragments of many unidentified beetles and the larvae of aquatic species.

Flies (Diptera) found in the pintail stomachs consisted mainly of larvae and amounted to 0.85 per cent of the total food. The following families of flies were represented: Midges (Chironomidae), in 31; brineflies (Ephydridae), in 15; soldierflies (Stratiomyidae), in 8; craneflies (Tipulidae), in 3; horseflies (Tabanidae), in 2; and unidentified fly remains in 10 stomachs.

The larvae or nymphs, and a few adults, of dragonflies (Anisoptera) and damselflies (Zygoptera), together furnished 0.44 per cent of the pintails' food; bugs (Heteroptera), consisting chiefly of water boatmen (Corixidae), creeping water bugs (Naucoridae), and giant water bugs (Belostomatidae), amounted to 0.23 per cent; the larvae and larval cases of caddisflies (Phryganoidea), 0.2 per cent; and other insects, consisting of a few grasshoppers (Orthoptera), ants and wasps (Hymenoptera), and Mayflies (Agnatha), totaled 0.2 per cent.

MISCELLANEOUS ANIMAL FOOD, 0.4 PER CENT.

The remains of small fish (found in 16 stomachs), a frog (*Rana* sp.), mandibles of a few marine worms (*Nereis* sp.), tiny water mites (Hydrachnidae), bits of hydroids (Hydrozoa), corallines (Bryozoa and Alcyonaria), and the minute, lime-incrusted, one-celled organisms known as Foraminifera, all were included in the varied bill of fare of the pintail.

WOOD DUCK.

Aix sponsa.

PLATE VII.

The wood duck ranges in summer nearly throughout the United States, southern British Columbia, southern Saskatchewan, Ontario, New Brunswick, and Nova Scotia; it breeds casually in Cuba, and is accidental in Bermuda, Mexico, and Jamaica. In winter it occupies approximately the southern half of its summer range. Its southward migration is accomplished chiefly in October, and it moves northward rather early in spring, reaching the latitude of central Iowa usually about March 20 and southern Manitoba April 15.

As its name implies, the wood duck inhabits secluded woodland ponds and lakes and timbered streams. It makes its nest in a natural cavity in a tree, from which the mother duck takes the young soon after they are hatched and carries them one by one in her beak to the water.[12] The adult male is the most brilliantly colored of all American ducks, and probably is as fine in appearance as any in the world. Its most striking feature is a large bright green and purple crest, striped with white, the rest of the head being of the same colors. The throat is white; on each side of the body is a row of black and white crescents, and across the shoulders are black and white bars. The upperparts are iridescent greenish or brownish black, and the breast is rich chestnut, spotted with white. The plumage of the female presents much the same general pattern as that of the male, but lacks most of its bright coloration.

FOOD HABITS.

Although the wood duck often is seen in the haunts of other ducks on open stretches of water or marshy land, its usual feeding grounds are along the banks of the wooded streams and ponds near which it nests in summer. Here it not only feeds upon the seeds and other parts of the plants which grow in or near the water, but often it wanders far out into the drier parts of the woods to pick up acorns, nuts, grapes, and berries, and the seeds of various trees and shrubs. Most of the insects and of the other animal food taken, however, are kinds which either inhabit the water itself or live on plants which grow in or near the water. Some terrestrial species are caught, but it is probable that most of these are picked up from the surface of the water, as the ducks are not fitted for successfully catching active insects on land. They are expert, however, in catching those which fly low over the water or glide over its surface, and obtain the kinds which swim beneath the surface (as well as

---

[12] Kingsford, E. G., Wood Duck Removing Young from the Nest: Auk, XXXIV., pp. 335-336, 1917.

the seeds and other parts of submerged plants) by half-diving, after the manner of the mallard and several other ducks.

The stomachs of 413[13] wood ducks were available for examination. Six stomachs were rejected on account of the too meager or uncertain nature of their contents, and seven more from ducks collected during January, May, and June, because they were not sufficient in number adequately to represent those months. None were taken during July; so that the 399 stomachs from which final results were computed represent only the months from August to December, and from February to April, inclusive. There is no reason why the food for January should not be similar to that of the other winter months; in summer, however, the percentage of animal food no doubt is somewhat higher. Stomachs were collected from 24 States and the District of Columbia, from Maine and Florida to Oregon and California, and from the Province of Ontario to Texas; but about five-eighths (268) of the whole number were taken in Louisiana. The bulk of the wood ducks, now nowhere abundant as breeders, inhabit the Mississippi Valley, and in winter they find ideal feeding and living conditions in the cypress swamps and wooded lakes and lagoons of the States bordering the Mississippi from about the mouth of the Ohio River southward.

VEGETABLE FOOD.

More than nine-tenths (90.19 per cent) of the food of the wood duck consists of vegetable matter. This high proportion of vegetable food is very similar to that taken by the mallard. With the wood duck it is quite evenly distributed among a large number of small items, chief among which are the following: Duckweeds, 10.35 per cent; cypress cones and galls, 9.25; sedge seeds and tubers, 9.14; grasses and grass seeds, 8.17; pondweeds and their seeds, 6.53; acorns and beechnuts, 6.28; seeds of waterlilies and leaves of water shield, 5.95; seeds of water elm and its allies, 4.75; of smartweeds and docks, 4.74; of coontail, 2.86; of arrow-arum and skunk cabbage, 2.42; of bur marigold and other composites, 2.38; of buttonbush and allied plants, 2.25; of bur reed, 1.96; wild celery and frogbit, 1.31; nuts of bitter pecan, 0.91; grape seeds, 0.82; and seeds of swamp privet and ash, 0.72 per cent. The remaining 9.4 per cent was made up of a large number of minor items.

DUCKWEEDS (LEMNACEAE), 10.35 PER CENT.

Whenever present in the feeding grounds of the wood duck, duckweeds probably are its favorite food. Each individual plant consists simply of a small fleshy leaf, disk shaped or nearly so, floating on the surface of the water, with one or more simple roots dangling.

---

[13] Eighty-six of these were examined by W. L. McAtee.

Bul. 862, U. S. Dept. of Agriculture. PLATE VII.

WOOD DUCK (AIX SPONSA).
Male on right; female on left.

Duckweeds are especially abundant on the still waters of southern cypress swamps, often covering the entire surface and furnishing an abundant supply of food for the ducks wintering there. The stomachs of many of the wood ducks taken in such localities in Louisiana and Missouri were filled almost entirely with duckweed plants, and the gullets also of several of them were crammed. A few ducks from other localities, as Arkansas, Illinois, New York, and Ontario, had taken this food in considerable quantities. Altogether, 99 of the wood ducks had been feeding upon greater duckweed (*Spirodela polyrhiza*) and 187 on other duckweeds (*Lemna* spp.).

PINE FAMILY (PINACEAE), 9.25 PER CENT.

The pine family was represented in the wood duck stomachs entirely by cone scales and galls from the bald cypress (*Taxodium distichum*), with possibly a few from pond cypress (*T. ascendens*). This peculiar diet is indulged in by this duck to a much greater extent than by any other, or probably by any other bird. The cones of cypress are about an inch in diameter, compact and nearly spherical, and when fully mature break up into angular woody scales, each containing a seed. It is these scales which the ducks pick up and which when ground by the powerful gizzards yield a starchy food material in the seeds. Several kinds of insect galls found on different parts of cypress trees also were eaten by the ducks. The kind most commonly taken was a hard, spherical gall made in the cone by a species of cecidomyid fly (*Retinodiplosis taxodii*). Cypress galls of various kinds were found in 35 of the stomachs, while 183 contained cone scales, some to the extent of 100 per cent of the contents.

SEDGES (CYPERACEAE), 9.14 PER CENT.

Sedge seeds are common articles of food of the wood duck, though not so much so as of most of the ducks which inhabit open marshes. A species of bulrush (*Scirpus cubensis*) which grows in swamps in the Gulf States far outweighed in importance any of the other sedges identified in the stomachs examined. The seeds of this plant were found in 47 stomachs, in several instances from 3,000 to more than 5,000 being present. Fifteen wood ducks had eaten the large, beaked seeds of pollywog or beaked rush (*Rhynchospora corniculata*). In a series of 3 stomachs from Okefenokee Swamp, Georgia, these seeds constituted 10, 17, and 35 per cent, respectively, of the contents. The small, hard, spherical seeds of saw grass (*Cladium effusum*) were present in 15 stomachs, those of nut rush (*Scleria* sp.) in 3. Seeds of chufas (*Cyperus* spp.) were found in 45 stomachs, usually in small numbers, and in one stomach from Minnesota was one large tuber of chufa (*Cyperus esculentus*). Seeds of the genus

Carex were commonly taken, usually in small numbers; unidentified kinds were present in 33 gizzards, the seeds of panicled sedge (*Carex decomposita*) in 21, and hop sedge (*C. lupuliformis*) in 8. Other sedge seeds found were those of river bulrush (*Scirpus fluviatilis*), *Fimbristylis*, spike rush (*Eleocharis* sp.), beaked rush (*Rhynchospora* sp.), and unidentified sedges.

GRASSES (GRAMINEAE), 8.17 PER CENT.

Chief among the grasses which contribute to the food of the wood duck is wild rice (*Zizania palustris*). Its seeds are fed upon by practically all the fresh water ducks, and it presents such an attractive source of food supply as to entice even the wood duck from its secluded haunts to the open marshes where the wild rice grows. According to Kumlien and Hollister,[14] the wood duck, in fall, "resorts to the great wild rice marshes, and while the rice lasts that seems to be its principal food." Wild rice had been eaten by 17 of the wood ducks examined, and when present usually furnished the bulk of the food. The stomach and gullet of one duck shot at Point Pelee, Ontario, contained no fewer than 1,200 seeds of wild rice, with remains of others. Another wood duck, taken at Sand Point, Mich., in August, had filled its craw and gizzard with about 400 flowers of the plant, whole heads of which had been swallowed. An article of food which seems to be very much relished by the ducks wherever found is the seeds of meadow grass (*Panicularia nervata*). Of a series of 22 wood ducks taken at Caruthersville, in extreme southeastern Missouri, 17 had been feeding upon these seeds. The stomachs and gullets of 7 contained, respectively, from 5,300 to 10,000 seeds to each individual, the seeds constituting from 75 to 96 per cent of the food. The seeds of a switchgrass (*Panicum*, subgenus *Dichanthelium*) were found in considerable quantities in several of the stomachs from Louisiana, some of which contained in addition remains of the stems and leaves of the same grass. In addition to the grasses already mentioned, the seeds of other switchgrasses (*Panicum* spp.), wild millet (*Echinochloa crus-galli*), cut-grass (*Zizaniopsis miliacea*), rice cut-grass (*Homalocenchrus oryzoides*), and love grass (*Eragrostis* sp.) were identified. The stomach of one wood duck taken near Chicago, Ill., in October, was filled with corn, and one from Louisiana contained traces of cultivated rice.

PONDWEEDS (NAIADACEAE), 6.53 PER CENT.

In the pondweeds is another family of plants which is important as a source of food for many species of ducks and to secure which the wood duck departs to some extent from its normal feeding habits.

---

[14] Kumlien, L., and N. Hollister, The Birds of Wisconsin: Bull. Wisconsin Nat. Hist. Soc., III, p. 21, 1903.

Seeds or other parts of five species of true pondweeds (*Potamogeton americanus*, *P. natans*, *P. heterophyllus*, *P. zosterifolius*, and *P. pectinatus*) each were found in from one to three stomachs, and unidentified pondweeds in 44. The gullet and stomach of one wood duck taken at Rush Lake, Michigan, in August, 1908, contained about 350 tubers of a species of *Potamogeton*. The gizzards of three from Delavan Lake, Wisconsin, were filled with the winter buds of eelgrass pondweed (*Potamogeton zosterifolius*). Only one wood duck had eaten seeds of bushy pondweed (*Najas flexilis*), and the seeds and leaves of widgeon grass (*Ruppia maritima*), which form so important an element of food for the gadwall and widgeon, were entirely lacking.

BEECH FAMILY (FAGACEAE), 6.28 PER CENT.

Acorns and beech nuts furnish one of the most important items of the wood duck's food, and had the collection of stomachs available been from localities more evenly distributed throughout its range, the percentage of this food very probably would have been much larger. As it was, 3 of the gizzards contained beech nuts and 19 contained acorns. Of the latter, 5 species were identified: Red oak (*Quercus rubra*) from 1 stomach, pin oak (*Q. palustris*) from 5, water oak (*Q. nigra*) from 2, black-jack (*Q. marylandica*) from 2, and valley oak (*Q. lobata*) from 1, while fragments of acorns found in 8 gizzards were not identified. Several of the stomachs containing acorns were crammed with them, the gullet and gizzard of one from Arkansas containing 15 entire acorns of pin oak, with fragments of one or two others. The wood duck's habit of eating acorns is well known, many writers testifying to its fondness for this kind of food. According to the late D. G. Elliot,[15] the wood duck is called "acorn duck" in Louisiana. Wilson and Bonaparte,[16] in 1831, wrote that its food "consists principally of acorns, seeds of the wild oats, and insects." Kumlien and Hollister[17] state that the wood duck "takes to the oak groves about the streams and lakes, and seems to be especially partial to the acorns of the bur oak. These it eats in large quantities."

Without doubt the wood duck's usual method of gathering acorns is by picking them up off the ground or from the water. However, one author, Mr. N. S. Goss,[18] is authority for the statement that, "Their food consists chiefly of insect life, the tender shoots and seeds of aquatic plants, grains, wild grapes and acorns, which they gather as well from the vines and tree tops as upon the ground. . . ."

[15] Elliot, D. G., Wild Fowl of North America, p. 87, 1898.
[16] Wilson, A., and C. L. Bonaparte, Amer. Ornith., III, p. 203, 1831.
[17] Kumlien, L., and N. Hollister, op. cit., III, p. 21, 1903.
[18] Goss, N. S., History of the Birds of Kansas, p. 73, 1891.

WATERLILY FAMILY (NYMPHAEACEAE), 5.95 PER CENT.

Seeds of waterlilies are quite frequently eaten by ducks. Twelve of the wood ducks had eaten seeds of yellow pondlilies, of which two species, the cowlily (*Nymphaea advena*) and the small yellow pondlily (*Nymphaea microphylla*) were identified; 10 contained seeds of white waterlilies, of which 2 species (*Castalia odorata* and *C. tuberosa*) were identified; and 11 contained seeds of water shield (*Brasenia schreberi*). Of the latter, one contained 380 seeds; and the stomach and distended gullet of a wood duck taken near Chicago, Ill., held 577 seeds of a white waterlily (*Castalia tuberosa*), with fragments of several more, in addition to other items. Two stomachs from southeastern Missouri contained quantities of the remains of stems and leaves of Carolina water shield (*Cabomba caroliniana*).

NETTLE FAMILY (URTICACEAE), 4.75 PER CENT.

The nettle family of plants was represented in the wood duck stomachs chiefly by the seeds of water elm (*Planera aquatica*),[19] which had been taken by 66 of these ducks. Many ducks in Louisiana had gorged themselves upon these large seeds, their gullets and gizzards being crammed. One wood duck had devoured nearly 300 at its last meal. A series of 13 gizzards from Avoyelles Parish, La., contained seeds of water elm to the average extent of over 48 per cent, most of the remainder of the food consisting of the seeds of coontail (*Ceratophyllum demersum*). The stomach of a duck from southeastern Missouri was nearly filled with the seeds of another elm (*Ulmus* sp.). Another, from Alabama, contained the remains of several red mulberries and eight of the hard drupes of hackberries (*Celtis* sp.), both of which belong to this family of plants. The small seeds of false nettle (*Boehmeria cylindrica*) were present in three stomachs.

SMARTWEEDS (POLYGONACEAE), 4.74 PER CENT.

In addition to a few seeds of dock (*Rumex* sp.), which were found in seven gizzards, and unidentified smartweeds, in 17, ten species of smartweed seeds were identified as having been eaten by the wood ducks. Of these, water smartweed (*Polygonum amphibium*) and mild water pepper (*P. hydropiperoides*) were the most commonly taken. The largest number of smartweed seeds found in any one wood duck stomach was taken from a bird collected in Avoyelles Parish, La., which had eaten over 1,100 of the small black seeds of *Polygonum opelousanum*.

COONTAIL (*Ceratophyllum demersum*), 2.86 PER CENT.

Coontail is a rootless, submerged plant, the much-branched stems of which bear bushy masses of small whorled leaves. Near the ends

---

[19] For a full description of this plant and its seeds, see Bull. 205, U. S. Dept. Agr., pp. 9–12, 1915.

of the branches are borne the fruits, consisting of hard, oblong, flattened seeds a little more than one-eighth of an inch long, and usually bearing on their outer covering from one to three woody spines. A few ducks, especially the gadwall and widgeon, relish the foliage of coontail, but most species, including the wood duck, prefer the seeds. Only a very few stomachs of this species contained foliage of coontail, and in these cases it was present in such small quantities as to indicate that it had been taken accidentally. The plant has a very wide distribution, and is found throughout North America, except in the extreme north. The seeds were found most commonly, however, in the stomachs of wood ducks from the Southern States. In a series of 65 gizzards from Moreauville, La., all but two of which contained seeds of coontail, they averaged about $12\frac{1}{2}$ per cent of the total food. In another series, 13 in number, from Avoyelles Parish, La., seeds of coontail amounted to more than 38 per cent. The largest number found in one gizzard was 127, with fragments of others.

ARUM FAMILY (ARACEAE), 2.42 PER CENT.

The large, starchy seeds of arrow-arum (*Peltandra virginica*) were present in 5 of the wood duck stomachs examined. Three of these from Portage Lake, Michigan, were well filled, one containing 51 seeds and remains of others, both crop and gizzard being crammed. In some localities these seeds are a very important item in the wood duck's food. In January, 1914, Lord William Percy found wood ducks in the Everglades of Florida feeding almost exclusively on the seeds of *Peltandra*. C. P. Alexander, writing of a visit to the Kinloch Gun Club, South Carolina, September 5, 1915, says: "As we approached, * * * several hundred summer ducks were feeding and flew up in small groups of 2 to 12. Upon examining the places from which they arose I found thousands of the seeds of *Peltandra* all neatly shelled out and the outer coats floating in small groups in the water. The spathes from which they were taken occurred by the score, each with a large hole torn in the side. Two of the ducks were shot and the craws were full of *Peltandra* seed. There can be no doubt of the importance of this plant as a food for *Aix* at least." The gullet and stomach of one wood duck from Connecticut contained 31 seeds of the skunk cabbage (*Symplocarpus foetidus*), with remains of several more.

COMPOSITES (COMPOSITAE), 2.38 PER CENT.

The flat, spined seeds of bur marigold (*Bidens* sp.), known as beggar-ticks or stick-tights, were present in 25 gizzards, sometimes in considerable numbers; three from Avoyelles Parish, La., were nearly filled with them. Two ducks collected in northeastern Kansas had filled up on the seeds of giant ragweed (*Ambrosia trifida*).

MADDER FAMILY (RUBIACEAE), 2.25 PER CENT.

One of the staple articles of food of the ducks feeding in the southern swamps is the seeds of buttonbush (*Cephalanthus occidentalis*). Of the total of 413 stomachs of wood ducks, 192 contained these seeds. They were present in 65 per cent of the stomachs from Louisiana, usually in small numbers. Occasionally, however, several hundred were present in single stomachs, in which they made as much as 75 or even 90 per cent of the contents. Remains of the seeds of buttonweed (*Diodia virginiana*) were found in 6 stomachs; those of cleavers (*Galium* sp.), in one.

BUR REEDS (SPARGANIACEAE), 1.96 PER CENT.

The hard, nutlike seeds of bur reeds (*Sparganium* spp.) were present in 53 of the wood duck stomachs. Bur reeds are aquatic plants with ribbon-shaped leaves and with seeds borne in clusters resembling burs at the ends of the branches. When these seeds are found in duck gizzards the outer coverings are usually worn off, but the hard kernels persist for some time. In 17 instances the seeds were identified as those of the broad-fruited bur reed (*Sparganium eurycarpum*).

FROGBIT FAMILY (HYDROCHARITACEAE), 1.31 PER CENT.

Remains of the many-seeded fruits of frogbit were found in 12 wood ducks' stomachs. Several full stomachs from Louisiana contained this food to the extent of from 65 to 90 per cent of their contents. The plant is aquatic, and its berries are fed upon by many species of ducks. Four wood ducks had been feeding upon wild celery (*Vallisneria spiralis*). One of these, taken from the shallows at the mouth of the Susquehanna River, in Maryland, had swallowed 30 of the sprouting winter buds of the plant, and another from Delavan, Wis., had filled up on the same food. This plant is a much more important element in the food of some of the deepwater ducks, as the red-head and the canvas-back, which obtain the winter buds and rootstocks by diving.

WALNUT FAMILY (JUGLANDACEAE), 0.91 PER CENT.

The powerful crushing and grinding ability of the wood duck's gizzard is shown by the presence in 76 of the stomachs examined of fragments of the nuts of the bitter pecan (*Hicoria aquatica*). These nuts have as hard a shell as any of the northern hickory nuts, yet they are broken as they enter the gizzard and before they can possibly have been exposed to the full crushing power of that organ.

All of the bitter pecans identified were from ducks taken in Louisiana. The entire food of one bird from Mansura, La., consisted of one whole pecan with fragments of several others. Usually pecans amounted to from 5 to 20 per cent of the stomach contents.

VINE FAMILY (VITACEAE), 0.82 PER CENT.

The estimated average percentage, 0.82, of the remains of grapes actually found in the wood duck stomachs, undoubtedly is much less than the true proportion of grapes consumed, for the skins, pulp, and juice are very quickly digested, leaving nothing but the seeds, or much more commonly, fragments of seeds, to show that grapes had been eaten. It is probable that most of these grapes are picked up from the ground in the woods, though some may be taken from the vines. Traces of grapes (always in the form of seeds or seed-fragments) were present in 141 of the wood duck stomachs.

OLIVE FAMILY (OLEACEAE), 0.72 PER CENT.

Two wood ducks had eaten seeds of ash (*Fraxinus americana* and other species). The remainder of the food from this family of plants consisted of the seeds of swamp privet (*Adelia acuminata*), which were present in 31 stomachs. This is a favorite food for wild ducks in some southern localities, according to the testimony of numerous hunters. "Wood ducks in particular are said to feed extensively upon its seeds. Weeks before other species of ducks arrive these birds are abundant in the country where swamp privet grows and are said to consume most of the crop of seeds, leaving little for other ducks."[20] The plant is a shrub or small tree, and the seed, which has a fibrous, ridged coat, is inclosed in a watery blue berry from one-half to three-fourths of an inch in length. These berries ripen in May and June and fall into the water; many of them are picked up from the bottom by the ducks later in the season. The swamp privet grows in the same kind of localities as the water elm, and its seeds usually were found in company with the seeds of that plant in the stomachs of wood ducks. One stomach and distended crop were found which held 157 of these large seeds, with remains of several more. A sprig of swamp privet was sent to the Biological Survey by C. G. Wright, of Dallas, Tex., with the statement that it was grown from seed taken from a wood duck's gizzard, which was absolutely full of them.

MISCELLANEOUS VEGETABLE FOOD, 9.4 PER CENT.

A great number of smaller items, each of which amounted to less than 1 per cent, made up the remainder of the wood duck's vegetable food. In some localities the ducks had fed upon the tubers and seeds of arrowheads (*Sagittaria latifolia* and other species); and three from Caruthersville, Mo., had stuffed themselves with the stems, leaves, and rootstocks of a crowfoot (*Ranunculus* sp.). Three taken in Okefenokee Swamp, Georgia, in December, 1916, had been feeding upon the crimson rootstocks of red-root or paint-root (*Gyrotheca*

---

[20] McAtee, W. L., Eleven Important Wild-duck Foods, Bull. No. 205, U. S. Dept. of Agr., pp. 12–13, 1915.

tinctoria) to the extent of 50, 60, and 70 per cent, respectively, of their stomach contents. These roots are said to be fed upon quite extensively by other waterfowl in that locality.

Several stomachs from southeastern Missouri contained from 1,000 to nearly 10,000 seeds each of lizard's-tail (*Saururus cernuus*), while others, chiefly from Louisiana swamps, held thousands of the small seeds of the primrose willow (*Jussiaea* sp.). The flat seeds of waterpenny (*Hydrocotyle* sp.) were present in 44 stomachs, but never in large numbers. Among the other aquatic plants whose seeds were commonly taken were the water milfoils (*Myriophyllum* sp.; *Proserpinaca* sp., and *Hippuris vulgaris*), pickerel weed (*Pontederia cordata*), fog-fruit (*Lippia* sp.), and swamp loosestrife (*Decodon verticillatus*). Seeds of wild heliotrope (*Heliotropium indicum*) and croton (*Croton* sp.) were found in 22 and 16 stomachs, respectively.

Many woodland shrubs, trees, and vines in addition to those already mentioned were represented by their seeds in a few stomachs each, or by only a few seeds in many stomachs. These included hollies (*Ilex* spp.), sumachs (*Rhus* spp.), supple jack (*Berchemia scandens*), buckthorn (*Rhamnus cathartica*), tupelo (*Nyssa sylvatica* and *N. aquatica*), dogwoods (*Cornus* spp.), storax (*Styrax* sp.), myrtles (*Myrica cerifera* and other species), hawthorns (*Crataegus* spp.), brambles (*Rubus* spp.), hornbeam (*Carpinus caroliniana*), sweet gum (*Liquidamber styraciflua*), greenbriar (*Smilax* sp.), and a few others.

ANIMAL FOOD.

The wood duck's animal food, which amounted to 9.81 per cent of the total, consisted chiefly of the following items: Dragonflies and damselflies and their nymphs, 2.54 per cent; bugs, 1.56; beetles, 1.02; grasshoppers and crickets, 0.23; flies and ants, bees, and wasps, 0.07; miscellaneous insects, 0.97; spiders and mites, 0.63; crustaceans, 0.08; and miscellaneous animal matter, 2.71 per cent. Thus, nearly two-thirds of the animal food consisted of insects.

DRAGONFLIES (ANISOPTERA); AND DAMSELFLIES (ZYGOPTERA), 2.54 PER CENT.

The food of 72 wood ducks included the remains of dragonflies and damselflies or their nymphs. The nymphs or larvae are much more commonly taken than the adults, as they are easier to catch. The food from this group of insects averaged 10.44 per cent of the total for 9 wood ducks taken in April, and 8.75 per cent for the 16 in March. During the remainder of the year much smaller quantities were taken.

BUGS (HETEROPTERA AND HOMOPTERA), 1.56 PER CENT.

Bugs, chiefly aquatic, are a very constant item of food for the wood duck, at least 17 families being represented in the contents of the stomachs examined. Of these the most important were the creeping water bugs (Naucoridae), of which the small, flat, predacious bugs of the genus *Pelocoris* had been taken by 43 of the ducks; the giant water bugs (Belostomatidae), represented by the genus *Belostoma* in 37 stomachs; water striders (Gerridae), found in 30; back-swimmers (Notonectidae), in 20; and water boatmen (Corixidae), in 15. Bugs of the last three families mentioned are without exception very active in their movements on or in the water, and their presence in so many stomachs of the wood duck no doubt is accounted for by their great abundance in lakes and rivers throughout most of North America. One wood duck taken at Alden, Wis., in August, 1908, had been feeding upon a species of plant louse (*Rhopalosiphum nymphaeae*) which in certain states of its development inhabits the leaves of waterlilies. The bird's gizzard contained about 1,600 of these plant lice, as well as other insects and seeds.

BEETLES (COLEOPTERA), 1.02 PER CENT.

Beetles of at least 15 families were represented in the food of the wood ducks examined. Of these the water scavenger beetles (Hydrophilidae), predacious diving beetles (Dytiscidae), and leaf beetles (Chrysomelidae) were most commonly taken. The first two families mentioned, as their names imply, are strictly aquatic, while the third was represented almost entirely by beetles of the genus *Donacia*, many of which feed upon aquatic plants, such as the pondlily, spatterdock, etc. Twenty-three of these beetles (*Donacia cincticornis*) were found in one stomach, together with a large number of seeds of the tuberous white waterlily (*Castalia tuberosa*), the plant on which they probably were captured. Two other strictly aquatic families, the whirligig beetles (Gyrinidae) and crawling water beetles (Haliplidae) were well represented. Six genera of ground beetles (Carabidae) were identified, and the leaf chafers (Scarabaeidae), long-horned beetles (Cerambycidae), and snout beetles (Curculionidae) had been eaten in considerable numbers. The fact that scarabaeid beetles have been eaten is often detected by the presence of small, hard grinding plates from their jaws, which frequently persist in bird stomachs long after all other parts of the beetles have been digested. Peculiar little silken cases containing eggs of water scavenger beetles, usually attached to a submerged leaf or to the body of the female beetle herself, are not infrequently found in duck stomachs. In two from Louisiana they made up 70 and 77 per cent, respectively, of the total contents.

GRASSHOPPERS, CRICKETS, ETC. (ORTHOPTERA), 0.23 PER CENT.

Grasshoppers of the genus *Orchelimum* were found in the stomachs of 11 wood ducks from Missouri; 39 mandibles, representing at least 20 grasshoppers, were present in one stomach; 8 wood ducks had eaten grouse locusts (Tettiginae).

FLIES (DIPTERA); AND ANTS, BEES, AND WASPS (HYMENOPTERA); 0.07 PER CENT.

Most of the flies eaten by wood ducks were in the larval form. Larvae of soldierflies (Stratiomyidae) predominated, having been taken by 33 of the ducks, while the gnats and midges (Chironomidae), craneflies (Tipulidae), flowerflies (Syrphidae), and the Scatophagidae were sparingly represented. Ants were commonly eaten, but always singly or in small numbers, never constituting a very large percentage of the total food. Seven genera were identified, *Crematogaster* and *Camponotus* predominating. Wasps (Vespoidea), parasitic wasps (Ichneumonoidea), sawflies (Tenthredinoidea), bees (Apoidea), chalcids (Chalcidoidea), and serphoids (Serphoidea) also were occasionally taken.

MISCELLANEOUS INSECTS, 0.97 PER CENT.

Prominent among the miscellaneous insect food of the wood ducks were the caterpillars and chrysalides of moths and butterflies (Lepidoptera), and the larvae and larval cases of caddisflies (Phryganoidea). One stomach examined contained no fewer than 85 noctuid moths, many of them with eggs. This family includes the cutworm moths, and practically all of its members are injurious to cultivated crops. A few Mayfly nymphs (Agnatha), termites (Isoptera), unidentified larvae, pupae, and galls, etc., were included among the miscellaneous insects.

SPIDERS AND MITES (ARACHNIDA), 0.63 PER CENT.

The wood duck's taste for spiders is quite marked. Several full stomachs from southern localities contained remains of from 20 to 40 spiders, these in some cases constituting as much as 75 or 80 per cent of the contents. Leathery or silken cases containing spider eggs occasionally are taken, and tiny water mites (Hydrachnidae) were found in 5 stomachs.

MISCELLANEOUS ANIMAL FOOD, 2.79 PER CENT.

Contrary to the habit of most other ducks, the wood duck pays little attention to mollusks, probably because they are not plentiful in its usual haunts. A very few snails and small bivalves were found in the stomachs examined. Crustaceans (0.08 per cent) were also scarce, being represented by a few beach fleas (Amphipoda), sowbugs (Oniscidae), water asels (Asellidae), and occasional claws of crawfish (Astacidae). Remains of small fishes, found in 5 stomachs; bones of frogs, in 2; 2 centipedes, and a few of the reproductive buds, or statoblasts, of fresh-water bryozoa, complete the list of food items.

FOOD HABITS OF SHOAL-WATER DUCKS.

TABLE I.—*Items of vegetable food identified in the stomachs of the ducks treated in this bulletin and the number of stomachs in which found.*

Kind of food.	Gad-wall.	Bald-pate.	Green-winged teal.	Blue-winged teal.	Cinna-mon teal.	Pin-tail.	Wood duck.
Total number of stomachs examined	417	255	653	319	41	790	413
SUBKINGDOM **EUTHALLOPHYTA**.							
Unidentified algae	29	8	7	4		8	4
Oedogonium sp		1					
Spirogyra sp	3	2					
Chara sp. (musk grass)	22	13	89	31	1	49	1
Nitella sp		1					
Unidentified lichens							1
SUBKINGDOM **PTERIDOPHYTA**.							
Marsileaceae.							
Marsilea sp. (pepperwort)						2	
Equisetaceae.							
Equisetum sp. (horsetail)		1	1				
SUBKINGDOM **SPERMATOPHYTA**.							
Pinaceae.							
Picea sp. (spruce), needles						21	
Taxodium distichum (bald cypress), cone scales	6		5				181
Taxodium distichum (bald cypress), galls							35
Taxodium sp. (cypress)							2
Sparganiaceae.							
Sparganium eurycarpum (bur reed)							17
Sparganium androcladum (bur reed)				2			1
Sparganium sp. (bur reed)	2	10	42	39	1	32	35
Naiadaceae.							
Unidentified	8						
Potamogeton natans (floating pondweed)				1			1
Potamogeton americanus (long-leaved pondweed)							1
Potamogeton heterophyllus (variable pondweed)							1
Potamogeton perfoliatus (curly pondweed)			1				
Potamogeton zosterifolius (eelgrass pondweed)							3
Potamogeton fresii (Fries pondweed)	1						
Potamogeton pusillus (small pondweed)	3	3	4			3	
Potamogeton diversifolius (curly pondweed)						1	
Potamogeton foliosus (leafy pondweed)						1	
Potamogeton pectinatus (sago pondweed)	9	5	7		1	10	2
Potamogeton sp. (unidentified pondweeds)	142	94	248	150	32	271	44
Ruppia maritima (widgeon grass)	112	92	111	87	16	254	
Zannichellia palustris (horned pondweed)	20	8	10	2	10	18	
Zostera marina (eelgrass)	3	10	3	3		93	
Najas marina (large bushy pondweed)				1			
Najas flexilis (bushy pondweed)	17	9	27	17		23	1
Juncaginaceae.							
Triglochin maritima (arrow-grass)		2	19			88	1
Alismaceae.							
Unidentified						2	2
Sagittaria latifolia (wapato)				1			2
Sagittaria teres (slender arrowhead)						2	
Sagittaria platyphylla (delta potato)	45					76	
Sagittaria sp. (unidentified arrowheads)		2	2	4		4	5

TABLE I.—*Items of vegetable food identified in the stomachs of the ducks treated in this bulletin and the number of stomachs in which found*—Continued.

Kind of food.	Gad-wall.	Bald-pate.	Green-winged teal.	Blue-winged teal.	Cinnamon teal.	Pin-tail.	Wood duck.
Total number of stomachs examined	417	255	653	319	41	790	413
SUBKINGDOM **SPERMATOPHYTA**—Continued.							
Alismaceae—Continued.							
Alisma plantago-aquatica (water plantain)			3	1		3	
Alisma subcordatum (American water plantain)			1				
Alisma sp. (water plantain)			1	1		3	
Hydrocharitaceae.							
Philotria canadensis (waterweed)		1					
Philotria sp. (waterweed)	1			2			
Vallisneria spiralis (wild celery)	3	12		1		4	4
Vallisneria sp. (wild celery)			2	1			
Limnobium spongia (frogbit)			3			4	12
Gramineae.							
Unidentified grasses	5	16	37	24	3	32	15
Paspalum sp.		1	4			2	
Syntherisma sanguinalis (crab grass)	1						
Panicum capillare (tumble weed)						1	
Panicum repens	4						
Panicum subgenus *Dichanthelium*							19
Panicum sp. (switch grass)	9	11	59	18		25	2
Panicum obtusum (range grass)		1					
Echinochloa crus-galli (wild millet)	1		14	1		9	1
Echinochloa colona (jungle rice)			1				
Echinochloa sp. (cockspur grass)				1		1	
Chaetochloa glauca (foxtail)	1		3	5		3	
Chaetochloa viridis (green foxtail)				1		2	
Chaetochloa sp. (foxtail)			6	8		23	
Zizania palustris (wild rice)		5	18	22		8	17
Zizaniopsis miliacea (cut-grass)	1		8	1		6	2
Homalocenchrus oryzoides (rice cut-grass)	1		3	9		4	4
Oryza sativa (cultivated rice)	19	1	21	12		52	1
Spartina glabra, var. *pilosa* (salt-marsh grass)						1	
Spartina sp. (salt-marsh grass)	1		1	1		4	
Phleum alpinum (mountain timothy)			1				
Sporobolus sp. (rush-grass)				1			
Agrostis sp. (bent grass)			1				
Beckmannia erucaeformis (Beckmann grass)			1				
Cynodon dactylon (Bermuda grass)						1	
Distichlis spicata (salt grass)	3	5	3			33	
Monanthochloë littoralis	1		16	13	5	6	
Eragrostis sp. (love grass)							1
Poa sp. (spear grass)	5					1	
Panicularia nervata (meadow grass)							17
Panicularia sp. (meadow grass)	1						
Puccinellia nuttalliana				3			
Festuca sp. (fescue grass)						1	
Avena sativa (cultivated oats)						1	
Hordeum vulgare (cultivated barley)	1					3	
Hordeum pusillum	2	1					
Triticum vulgare (cultivated wheat)			1			3	
Zea mays (cultivated corn)	1		1	1		3	1

FOOD HABITS OF SHOAL-WATER DUCKS. 51

TABLE I.—*Items of vegetable food identified in the stomachs of the ducks treated in this bulletin and the number of stomachs in which found*—Continued.

Kind of food.	Gad-wall.	Bald-pate.	Green-winged teal.	Blue-winged teal.	Cinna-mon teal.	Pin-tail.	Wood duck.
Total number of stomachs examined	417	255	653	319	41	790	413
SUBKINGDOM **SPERMATOPHYTA**—Continued.							
Cyperaceæ.							
Unidentified sedges	43	20	50	27	7	47	5
Cyperus esculentus (chufa)							1
Cyperus sp. (chufa)	31	5	48	45		29	45
Dichromena sp			1				
Dulichium arundinaceum (three-ways sedge)				1			
Eleocharis sp. (spike rush)	13	19	45	33	10	45	2
Fimbristylis sp	5	4	90	40		61	8
Scirpus pauciflorus (few-flowered bulrush)						5	
Scirpus americanus (three-square)	150	37	121	10	3	155	
Scirpus paludosus (prairie bulrush)	27	12	46	7	12	24	
Scirpus robustus (salt-marsh bulrush)	24		40	2		24	
Scirpus fluviatilis (river bulrush)	2	5	5	18		3	4
Scirpus cubensis (bulrush)	2		13			8	47
Scirpus validus (great bulrush)			38	2			
Scirpus sp. (unidentified bulrushes)	47	24	205	184	17	154	16
Fuirena squarrosa (umbrella-grass)						1	
Fuirena sp. (umbrella-grass)						1	
Rhynchospora corniculata (beaked rush)			1			7	15
Rhynchospora sp. (beaked rush)			4	2		8	2
Cladium effusum (saw grass)	68	5	83	46		103	15
Cladium mariscoides (twig-rush)		1	8	9		1	
Scleria sp. (nut rush)						6	3
Carex decomposita (panicled sedge)	1		10	2		2	21
Carex lupuliformis (hop sedge)							8
Carex sp. (sedge)	8	12	62	57	5	19	33
Araceae.							
Peltandra virginica (arrow-arum)							5
Symplocarpus foetidus (skunk cabbage)							1
Lemnaceae.							
Spirodela polyrhiza (big duckweed)							99
Lemna perpusilla (minute duckweed)							1
Lemna minor (small duckweed)							1
Lemna sp. (unidentified duckweeds)	17	3	44	14		15	185
Eriocaulaceae.							
Eriocaulon decangulare (pipewort)							1
Eriocaulon sp. (pipewort)			1				
Pontederiaceae.							
Pontederia cordata (pickerel weed)	1		1			8	9
Heteranthera dubia (water star-grass)					1		2
Heteranthera sp. (mud plantain)						1	1
Juncaceae.							
Juncus sp. (bog rush)			2			2	
Liliaceae.							
Polygonatum biflorum (hairy Solomon's-seal)							1
Smilax sp. (greenbriar)							2
Haemodoraceae.							
Gyrotheca tinctoria (red-root)							3
Piperaceae.							
Saururus cernuus (lizard's-tail)							15

TABLE I.—*Items of vegetable food identified in the stomachs of the ducks treated in this bulletin and the number of stomachs in which found*—Continued.

Kind of food.	Gadwall.	Baldpate.	Greenwinged teal.	Bluewinged teal.	Cinnamon teal.	Pintail.	Wood duck.
Total number of stomachs examined	417	255	653	319	41	790	413
SUBKINGDOM **SPERMATOPHYTA**—Continued.							
Salicaceae.							
Salix sp. (willow), galls				1			
Salix sp. (willow), capsules	1						1
Myricaceae.							
Myrica cerifera (wax myrtle)							6
Myrica californica (California myrtle)						3	
Myrica sp. (unidentified myrtles)	7	5	4	18		42	3
Juglandaceae.							
Hicoria aquatica (bitter pecan)							76
Betulaceae.							
Carpinus caroliniana (hornbeam; blue beech)							5
Betula sp. (birch)			5				
Alnus sp. (alder)			2			1	
Fagaceae.							
Fagus grandifolia (American beech)							3
Quercus rubra (red oak)							1
Quercus palustris (pin oak)							5
Quercus nigra (water oak)							2
Quercus marylandica (black-jack oak)							2
Quercus lobata (valley oak)							1
Quercus sp. (unidentified acorns)							8
Urticaceae.							
Ulmus sp. (elm)							1
Planera aquatica (water elm)							66
Celtis sp. (hackberry)	1						1
Morus rubra (red mulberry)							1
Boehmeria cylindrica (false nettle)							3
Boehmeria sp. (false nettle)			2				
Polygonaceae.							
Rumex persicarioides (golden dock)			1				
Rumex sp. (dock)	2		4	2	2	7	7
Polygonum amphibium (water smartweed)	3	11	34	27	3	23	24
Polygonum arifolium (prickly smartweed)							1
Polygonum aviculare (knotweed)	2	2	8			7	
Polygonum convolvulus (black bindweed)	1	1	1			1	
Polygonum hydropiper (water pepper)	2	2	12	10		5	8
Polygonum hydropiperoides (mild water pepper)	2	1	10	31		29	24
Polygonum lapathifolium (dock-leaved smartweed)	5	6	29	26	3	6	4
Polygonum opelousanum (smartweed)	1		14	7		4	10
Polygonum pennsylvanicum (Pennsylvania smartweed)	2		3	3		5	4
Polygonum persicaria (lady's-thumb)	2	2	9	10	1	3	2
Polygonum portoricense (dense-flowered smartweed)			3	1			
Polygonum punctatum (dotted smartweed)			4			1	1
Polygonum sagittatum (arrow-leaved smartweed)	2		7	11		10	8
Polygonum sp. (unidentified smartweeds)			22	16	3	22	17

FOOD HABITS OF SHOAL-WATER DUCKS. 53

TABLE I.—*Items of vegetable food identified in the stomachs of the ducks treated in this bulletin and the number of stomachs in which found*—Continued.

Kind of food.	Gad-wall.	Bald-pate.	Green-winged teal.	Blue-winged teal.	Cinna-mon teal.	Pin-tail.	Wood duck.
Total number of stomachs examined	417	255	653	319	41	790	413
SUBKINGDOM **SPERMATOPHYTA**—Continued.							
Chenopodiaceae:							
Chenopodium album (lamb's-quarters)			3				
Chenopodium sp. (pigweed)	1			1	1	5	
Atriplex sp. (saltbush)		2	2			1	
Salicornia ambigua (glasswort; picklegrass)	1	3	1			53	
Amaranthaceae.							
Amaranthus retroflexus (green amaranth)			2				
Amaranthus sp. (pigweed)	1		20	5	1	15	
Caryophyllaceae.							
Spergula arvensis (corn spurrey)			2			2	
Arenaria sp			1				
Unidentified				1			
Portulacaceae.							
Portulaca oleracea (common purslane)						1	
Portulaca sp. (purslane)	2	2	3			1	
Ceratophyllaceae.							
Ceratophyllum demersum (coontail; hornwort)	59	3	20	7		25	186
Nymphaeaceae.							
Unidentified						1	
Nymphaea advena (cowlily; spatter-dock)							4
Nymphaea microphylla (small yellow pondlily)							5
Nymphaea mexicana (banana waterlily)						1	
Nymphaea sp. (yellow pondlily)		2	2			6	3
Castalia odorata (sweet-scented waterlily)						1	3
Castalia tuberosa (white waterlily)							4
Castalia sp. (waterlily)	3	1	10	14		9	3
Castalia sp., tubers			1				
Brasenia schreberi (water shield)	1	2	20	13		34	11
Cabomba caroliniana (Carolina water shield)							2
Ranunculaceae.							
Ranunculus delphinifolius (yellow water-crowfoot)			2	1	1		
Ranunculus sp. (crowfoot)	5	8	8	6		17	13
Papaveraceae			3				
Cruciferae.							
Unidentified						1	
Brassica sp. (mustard)			1				1
Barbarea sp. (winter cress)			1				
Hamamelidaceae.							
Liquidambar styraciflua (sweet gum)				7			5
Rosaceae.							
Unidentified						1	
Crataegus sp. (hawthorn)			2	1		12	6
Fragaria sp. (strawberry)				1			
Rubus sp. (bramble)	1	2	15	15		43	4
Rosa sp. (rose)						1	
Leguminosae.							
Cassia marylandica (American senna)						1	
Cassia sp. (unidentified senna)						7	
Trifolium arvense (rabbit's-foot clover)			1				
Trifolium sp. (clover)	1		6	3	1	1	

TABLE I.—*Items of vegetable food identified in the stomachs of the ducks treated in this bulletin and the number of stomachs in which found*—Continued.

Kind of food.	Gadwall.	Baldpate.	Greenwinged teal.	Bluewinged teal.	Cinnamon teal.	Pintail.	Wood duck.
Total number of stomachs examined.	417	255	658	819	41	790	413
SUBKINGDOM **SPERMATOPHYTA**—Continued.							
Leguminosae—Continued.							
Melilotus sp. (sweet clover).		2		3			
Medicago denticulata (bur clover).		1	1		2		
Medicago sp.					1		
Vicia sp. (vetch).						1	
Vigna sinensis (cow pea).						2	
Unidentified.			1				
Geraniaceae.							
Geranium sp. (cranesbill).			1				
Erodium sp. (storksbill).						1	
Euphorbiaceae.							
Croton texensis.				1			
Croton sp. (spurge).		1	1			2	16
Euphorbia sp.			1				1
Empetraceae.							
Empetrum nigrum (crowberry).			1				
Anacardiaceae.							
Rhus glabra (smooth sumach).							1
Rhus toxicodendron (poison ivy).	1		1				1
Rhus laurina (California sumach).					1		
Rhus sp. (unidentified sumachs).	25	1	1	2		1	24
Aquifoliaceae.							
Ilex sp. (holly).	5	1	1	5		7	41
Rhamnaceae.							
Berchemia scandens (supple jack).							13
Rhamnus cathartica (buckthorn).							1
Vitaceae.							
Vitis cordifolia (frost grape).							3
Vitis sp. (grapes).	2		22	2		11	138
Malvaceae.							
Unidentified.					2	3	
Abutilon abutilon (Indian mallow).						1	
Sida spinosa (nail grass).			1	2			
Sida sp. (nail grass).						3	
Malva sp. (mallow).				1		1	
Kosteletzkya virginica.						1	
Hibiscus sp. (rose mallow).				2		1	1
Hypericaceae.							
Hypericum sp. (St. John's-wort).							1
Cactaceae.							
Opuntia sp. (prickly pear).						5	
Lythraceae.							
Decodon verticillatus (swamp loosestrife; willow herb).			1	1			2
Onagraceae.							
Jussiaea sp. (primrose willow).				6			8
Haloragidaceae.							
Myriophyllum verticillatum (whorled milfoil).		1					
Myriophyllum sp. (water milfoil).	9	23	58	44	2	24	12
Proserpinaca sp. (mermaid weed).	1		1	5		6	18
Hippuris vulgaris (bottle brush).	6	18	16	8	3	12	2

FOOD HABITS OF SHOAL-WATER DUCKS.

TABLE I.—*Items of vegetable food identified in the stomachs of the ducks treated in this bulletin and the number of stomachs in which found*—Continued.

Kind of food.	Gadwall.	Baldpate.	Greenwinged teal.	Bluewinged teal.	Cinnamon teal.	Pintail.	Wood duck.
Total number of stomachs examined	417	255	653	319	41	790	413
SUBKINGDOM **SPERMATOPHYTA**—Continued.							
Umbelliferae.							
Apium sp. (wild parsley)			1				
Hydrocotyle sp. (water pennywort)	3		19	12		3	44
Centella asiatica (marsh pennywort)				2		7	
Cicuta curtisii (water hemlock)						1	
Cicuta sp. (water hemlock)		5				3	
Cornaceae.							
Cornus amomum (kinnikinnik)							3
Cornus asperifolia (rough-leaved dogwood)							3
Cornus sp. (dogwood)	1	1	2				1
Nyssa sylvatica (sour gum)							9
Nyssa aquatica (large tupelo)							2
Nyssa sp. (tupelo)						1	11
Ericaceae.							
Gaultheria sp. (wintergreen)				2			
Gaylussacia sp. (huckleberry)			2			1	
Vaccinium sp. (blueberry)		1	1			7	1
Styracaceae.							
Styrax grandifolia (storax)							1
Styrax sp. (storax)							7
Oleaceae.							
Fraxinus americana (white ash)							1
Fraxinus sp. (ash)							1
Adelia acuminata (swamp privet)							31
Gentianaceae.							
Menyanthes trifoliata (bog bean)			2			2	
Asclepiadaceae.							
Asclepias sp. (milkweed)		1				1	
Convolvulaceae.							
Cuscuta sp. (dodder)	2	2	19	11		8	1
Boraginaceae.							
Heliotropium indicum (wild heliotrope)	1	3	47	16	2	28	22
Verbenaceae.							
Verbena hastata (blue verbena)			1				
Verbena sp. (verbena)	1		3	11		1	1
Lippia nodiflora (fog-fruit)						1	
Lippia sp. (fog-fruit)		1	22	13		5	7
Callicarpa americana (French mulberry)						1	
Labiatae.							
Unidentified						1	
Trichostema sp. (blue curls)						1	
Mentha sp. (mint)			2				
Solanaceae.							
Solanum sp. (night shade)			5				
Scrophulariaceae.							
Monniera rotundifolia (water hyssop)			5			1	
Plantaginaceae.							
Plantago sp. (plantain)			5	1			

TABLE I.—*Items of vegetable food identified in the stomachs of the ducks treated in this bulletin and the number of stomachs in which found*—Continued.

Kind of food.	Gad-wall.	Bald-pate.	Green-winged teal.	Blue-winged teal.	Cinna-mon teal.	Pin-tail.	Wood duck.
Total number of stomachs examined	417	255	653	319	41	790	413
SUBKINGDOM **SPERMATOPHYTA**—Continued.							
Rubiaceae.							
Galium sp. (cleavers)	1	1	2	10	2	3	1
Diodia virginiana (buttonweed)							6
Diodia teres (rough buttonweed)				1		1	
Cephalanthus occidentalis (buttonbush)	23	2	29	5		8	192
Caprifoliaceae.							
Symphoricarpos sp. (snowberry)			2				1
Sambucus sp. (elder)		1	5	1		32	
Compositae.							
Unidentified	2	1					1
Ambrosia artemisifolia (ragweed)			1	3		1	
Ambrosia trifida (giant ragweed)						2	2
Ambrosia sp. (ragweed)			4	5		3	
Xanthium spinosum (cocklebur)			1				
Helianthus maximiliana (sunflower)			1				
Helianthus sp. (sunflower)	1	1	1			4	
Bidens sp. (bur marigold)	5	2	3	2		6	25
Carduus sp. (plumeless thistle)	1		1			3	
Taraxacum sp. (dandelion)				1			
Lactuca sp. (lettuce)			2				

TABLE II.—*Items of animal food identified in the stomachs of the ducks treated in this bulletin and number of stomachs in which found.*

Kind of food.	Gad-wall.	Bald-pate.	Green-winged teal.	Blue-winged teal.	Cinna-mon teal.	Pin-tail.	Wood duck.
Total number of stomachs examined	417	255	653	319	41	790	413
SUBKINGDOM **PROTOZOA**.							
Foraminifera			13			1	
SUBKINGDOM **COELENTERATA**.							
Hydrozoa (hydroids)	2	2	3			13	
Alcyonaria						1	
SUBKINGDOM **MOLLUSCOIDA**.							
Phylactolaemata (fresh-water Bryozoa)	7		2			1	2
SUBKINGDOM **ANNULATA**.							
Nereidae.							
Nereis sp. (marine worms)			3			5	
SUBKINGDOM **ARTHROPODA**.							
CLASS **Crustacea** (CRUSTACEANS).							
Unidentified	1	2	11	5		4	

FOOD HABITS OF SHOAL-WATER DUCKS.

TABLE II.—*Items of animal food identified in the stomachs of the ducks treated in this bulletin and number of stomachs in which found.*—Continued.

Kind of food.	Gadwall.	Baldpate.	Greenwinged teal.	Bluewinged teal.	Cinnamon teal.	Pintail.	Wood duck.
Total number of stomachs examined	417	255	653	319	41	790	413
CLASS Crustacea—Continued.							
Order OSTRACODA.							
Unidentified ostracods (bivalved crustaceans)	21	2	61	8	1	35	1
Cytheridae.							
Cythere sp						1	
Cyprididae.							
Candona sigmoides						1	
Candona simpsoni						1	
Order AMPHIPODA.							
Unidentified		2	11	4		17	5
Corophiidae.							
Corophium cylindricum			1				
Corophium sp			1				
Gammaridae.							
Pseudalibrotus littoralis		1					
Gammarus fasciatus (sand fleas)						2	
Gammarus locusta				1			
Gammarus sp		4	13				
Orchestiidae.							
Hyalella azteca							1
Hyalella knickerbockeri				1			
Hyalella dentata			1	1		1	
Podoceridae.							
Amphithoë longimana		1					
Order COPEPODA.							
Unidentified copepods (water fleas)			1				
Order ISOPODA.							
Unidentified			2			4	1
Oniscidae.							
Unidentified sowbugs	1					1	1
Asellidae.							
Asellus sp. (asel)							1
Mancasellus brachyurus (asel)							1
Order DECAPODA.							
Suborder MACRURA.							
Crangonidae.							
Unidentified shrimps						2	
Crangonyx gracilis (shrimp)				1			
Astacidae.							
Unidentified crawfishes	2					4	2
Cambarus propinquus							1
Cambarus sp							1
Suborder BRACHYURA.							
Unidentified crabs	2		1	1		18	
Pilumnidae.							
Hexapanopeus angustifrons						10	
Neopanope texana (Say crab)	1						

TABLE II.—*Items of animal food identified in the stomachs of the ducks treated in this bulletin and number of stomachs in which found*—Continued.

Kind of food.	Gadwall.	Baldpate.	Greenwinged teal.	Bluewinged teal.	Cinnamon teal.	Pintail.	Wood duck.
Total number of stomachs examined	417	255	653	319	41	790	413
CLASS **Myriapoda**.							
Unidentified centipedes			1				2
CLASS **Insecta** (INSECTS).							
Unidentified insect fragments, eggs, larvae, and pupae	9	13	52	14	1	23	24
Unidentified insect galls							5
Order THYSANURA.							
Unidentified							2
Superorder AMPHIBIOTICA.							
Unidentified damselflies or dragonflies and nymphs	1	1		3	1	5	32
Order ZYGOPTERA (Damselflies).							
Unidentified damselflies and nymphs	2		2			2	1
Agrionidae.							
Nehalennia sp			1				
Enallagma sp				2		1	
Order ANISOPTERA (Dragonflies).							
Unidentified dragonflies and nymphs	2		17	22	2	21	41
Aeschnidae, nymphs			1			3	1
Libellulidae.							
Unidentified nymphs			1				2
Somatochlora sp						1	
Sympetrum sp			1				
Cordulia sp						2	
Order AGNATHA (Mayflies).							
Unidentified Mayfly nymphs	1					1	2
Order PLECOPTERA (Stoneflies).							
Unidentified stonefly larvae			3				
Order ISOPTERA (Termites).							
Claotermes castaneus (termite)							1
Order ORTHOPTERA (Grasshoppers, etc.).							
Unidentified grasshoppers and their eggs			4	1		5	2
Acridiidae.							
Unidentified		1				1	
Melanoplus sp							1
Tettiginae (grouse locusts).							
Unidentified							7
Tettigidea sp							1
Locustidae (green grasshoppers).							
Unidentified							2
Orchelimum sp							11
Gryllidae (crickets).							
Nemobius sp		2					1

FOOD HABITS OF SHOAL-WATER DUCKS.

TABLE II.—*Items of animal food identified in the stomachs of the ducks treated in this bulletin and number of stomachs in which found*—Continued.

Kind of food.	Gadwall.	Baldpate.	Green-winged teal.	Blue-winged teal.	Cinnamon teal.	Pintail.	Wood duck.
Total number of stomachs examined	417	255	653	319	41	790	413
CLASS **Insecta**—Continued.							
Order MALLOPHAGA (Bird Lice).							
Unidentified			1				
Order HETEROPTERA (True Bugs).							
Unidentified bugs			6	2		3	15
Corixidae (water boatmen).							
Unidentified species	25	9				33	15
Corixa sp			32	43	11		
Belostomatidae (giant water bugs).							
Belostoma sp				1		1	37
Nepidae (water scorpions).							
Ranatra fusca							2
Naucoridae (creeping water bugs).							
Pelocoris femoratus				2		2	
Pelocoris sp	6			13		7	43
Notonectidae (back-swimmers).							
Notonecta sp				1			2
Buenoa sp							1
Plea striola			4	11			17
Saldidae (shorebugs).							
Salda sp	1						
Veliidae (broad-shouldered water striders).							
Microvelia sp							7
Velia australis							4
Gerridae (water striders).							
Gerris marginata							1
Gerris sp	4		4	2			29
Hydrometridae.							
Hydrometra sp							4
Miridae.							
Unidentified		1					
Lygus pratensis							1
Mesoveliidae.							
Mesovelia mulsanti				2			6
Hebridae.							
Hebrus sp							1
Reduviidae (assassin bugs)							2
Coreidae.							
Darmistus sp							1
Pentatomidae (stink bugs).							
Unidentified	1						7
Menecles incertus				1			
Euschistus variolarius							1
Euschistus sp							2
Corimelaenidae.							
Corimelaena nitiduloides				1			

TABLE II.—*Items of animal food identified in the stomachs of the ducks treated in this bulletin and number of stomachs in which found*—Continued.

Kind of food.	Gadwall.	Baldpate.	Greenwinged teal.	Bluewinged teal.	Cinnamon teal.	Pintail.	Wood duck.
Total number of stomachs examined	417	255	653	319	41	790	413
CLASS **Insecta**—Continued.							
Order HOMOPTERA.							
Cicadellidae.							
Draeculacephala mollipes							1
Fulgoridae.							
Unidentified				1			
Scolops sp	1						
Dicranotropis sp							1
Jassidae (leafhoppers)			1	1			
Aphididae (plant lice).							
Rhopalosiphum nymphaeae							4
Order MEGALOPTERA (Fishflies).							
Unidentified larvae							1
Order NEUROPTERA.							
Unidentified			1				
Sialidae (dobson, etc.)			1	1			
Order PHYRYGANOIDEA (Caddisflies).							
Unidentified larvae and cases	9	7	46	31	1	34	4
Phryganeidae.							
Phryganea improba							1
Limnephilidae.							
Stenophylax sp							1
Sericostomatidae.							
Brachycentrus incanus							1
Hydropsychidae.							
Hydropsyche sp							1
Order LEPIDOPTERA (Butterflies and Moths).							
Hemileucidae.							
Hemileuca maia							1
Arctidae							1
Noctuidae							1
Geometridae, pupa							1
Pyralidae							2
Tineidae, cocoon				3			
Unidentified moths			1				2
Unidentified pupae							6
Unidentified caterpillars	1	1					13
Order COLEOPTERA (Beetles).							
Unidentified fragments and larvae	18	16	40	50	21	40	62
Cicindelidae (tiger beetles)							1
Carabidae (ground beetles).							
Unidentified	2		6	3		3	6
Carabus vinctus							1
Scarites substriatus							1
Scarites sp						1	

FOOD HABITS OF SHOAL-WATER DUCKS.

TABLE II.—*Items of animal food identified in the stomachs of the ducks treated in this bulletin and number of stomachs in which found*—Continued.

Kind of food.	Gad-wall.	Bald-pate.	Green-winged teal.	Blue-winged teal.	Cinnamon teal.	Pintail.	Wood duck.
Total number of stomachs examined	417	255	653	319	41	790	413
CLASS **Insecta**—Continued.							
Order COLEOPTERA—Continued.							
Carabidae (ground beetles)—Continued.							
Aspidoglossa subangulata				1			
Bembidium intermedium				1			
Bembidium insulatum	1						
Bembidium sp	1					1	
Loxandrus velox						1	
Platynus sp				1			1
Chlaenius sp				1		2	
Anomoglossus pusillus							1
Harpalus caliginosus							1
Selenophorus sp				1			
Stenolophus conjunctus						1	
Anisodactylus dulcicollis				1			
Anisodactylus rusticus				1			
Haliplidae (crawling water beetles).							
Unidentified			4			1	
Haliplus triopsis							2
Haliplus unicolor				1			
Haliplus sp				2		1	1
Cnemidotus edentulus						1	
Cnedmidotus pedunculatus							1
Cnemidotus sp							1
Peltodytes simplex				1			
Peltodytes callosus					1		
Peltodytes sp				4			
Dytiscidae (predacious diving beetles).							
Unidentified adults and larvae	3	2	22	8	2	4	9
Colpius inflatus				3		3	1
Canthydrus bicolor				1		1	
Canthydrus puncticollis						1	
Hydrocanthus iricolor				1		2	
Hydrovatus compressus						1	
Hydrovatus sp							1
Bidessus affinis			1	2			
Bidessus obscurellus					1		
Bidessus flavicollis			1	1			
Bidessus sp				2		1	
Coelambus acaroides							1
Coelambus inaequalis						1	
Coelambus punctatus						1	
Coelambus turbidus				1			
Coelambus sp						2	1	1
Hydroporus sp				1			
Coptotomus interrogatus				1			
Coptotomus sp							2
Agabus sp				1		1	
Laccophilus decipiens			1		1		

TABLE II.—*Items of animal food identified in the stomachs of the ducks treated in this bulletin and number of stomachs in which found*—Continued.

Kind of food.	Gadwall.	Baldpate.	Greenwinged teal.	Bluewinged teal.	Cinnamon teal.	Pintail.	Wood duck.
Total number of stomachs examined	417	255	653	319	41	790	413
CLASS Insecta—Continued.							
Order COLEOPTERA—Continued.							
Gyrinidae (whirligig beetles).							
Unidentified							3
Gyrinus sp				1			3
Dineutes sp							3
Hydrophilidae (water scavenger beetles).							
Unidentified beetles and larvae	3		5	11	1	16	8
Unidentified egg-cases							3
Hydrophilus californicus					2		
Hydrophilus sp						2	2
Tropisternus sp							7
Helophorus sp				3	2		
Hydrocharis obtusatus						1	
Berosus pantherinus				2			
Berosus peregrinus						1	
Berosus striatus						1	
Berosus sp	4	7	7	6	1	8	4
Philhydrus perplexus						1	
Philhydrus nebulosus							2
Philhydrus sp				1			2
Hydrobius sp	1						
Cercyon sp		1					
Staphylinidae (rove beetles)	1	1	2			2	2
Coccinellidae (ladybugs).							
Unidentified	1						
Megilla maculata	1						
Hippodamia sp							1
Cucujidae (flat bark beetles).							
Silvanus surinamensis						1	
Dermestidae (larder beetles).							
Unidentified		1					
Dermestes lardarius		1					
Histeridae.							
Hister sp				1			
Nitidulidae.							
Stelidota geminata							1
Byrrhidae	1			1			
Parnidae.							
Elmis vittatus			1				
Heteroceridae.							
Heterocerus sp				2			
Elateridae						1	1
Lucanidae.							
Passalus cornutus							2
Scarabaeidae (leaf chafers).							
Unidentified			1			2	11
Onthophagus hecate				1			
Aphodius inquinatus		1				1	1
Aphodius sp	1	1	2	1			1
Phyllophaga sp. (May beetles)				1			2
Dyscinetus trachypygus							1

FOOD HABITS OF SHOAL-WATER DUCKS.

TABLE II.—*Items of animal food identified in the stomachs of the ducks treated in this bulletin and number of stomachs in which found—Continued.*

Kind of food.	Gadwall.	Baldpate.	Greenwinged teal.	Bluewinged teal.	Cinnamon teal.	Pintail.	Wood duck.
Total number of stomachs examined	417	255	653	319	41	790	413
CLASS **Insecta**—Continued.							
Order COLEOPTERA—Continued.							
Cerambycidæ (long-horned beetles).							
Unidentified							1
Leptostylus aculiferus							1
Liopus sp							1
Chrysomelidae (leaf beetles).							
Unidentified	1	2		2		2	
Donacia cincticornis							2
Donacia subtilis							2
Donacia proxima							1
Donacia sp	1	1	2	2	1	1	11
Xanthonia decemnotata	1						
Prasocuris phellandrii				1			
Ceratoma trifurcata							1
Luperodes meraca	1						
Galerucella sp	2					1	
Phyllotreta pusilla	1						
Phyllotreta sp				1			
Tenebrionidae	1						
Melandryidae.							
Synchroa punctata							1
Anthicidae (flower beetles).							
Anthicus haldemani	1						
Anthicus sp	1						
Meloidae (blister beetles)	1						
Suborder RHYNCHOPHORA (Weevils).							
Unidentified adults	12	1	17	8	2	14	12
Unidentified larvae							2
Curculionidae (snout beetles).							
Unidentified				1	2		
Otiorhynchinae						1	
Hypera punctata							3
Erirhinini						2	
Lixellus filiformis				1			
Baqous sp	1						
Endalus sp						1	
Sphenophorus aequalis				1			
Sphenophorus ochreus		1					
Sphenophorus sp				3		2	1
Calandra sp			1				
Order DIPTERA (Flies).							
Unidentified adults, larvae, and pupae	13	1	14	8	2	10	4
Tipulidae (craneflies).							
Unidentified adults, larvae, and pupae	1	3	5			3	
Ctenophorinae, larvae							1

TABLE II.—*Items of animal food identified in the stomachs of the ducks treated in this bulletin and number of stomachs in which found*—Continued.

Kind of food.	Gadwall.	Baldpate.	Greenwinged teal.	Bluewinged teal.	Cinnamon teal.	Pintail.	Wood duck.
Total number of stomachs examined	417	255	653	319	41	790	413
CLASS **Insecta**—Continued.							
Order DIPTERA—Continued.							
Chironomidae (midges).							
Unidentified adults, larvae, and pupae	10	12	61	10	2	31	2
Chironomus sp							1
Ceratopogon sp				1			
Simuliidae.							
Simulium sp			1	1			
Stratiomyidae (soldierflies).							
Unidentified larvae			3	11	3	7	31
Odontomyia sp	2					1	2
Nemotelus sp							1
Tabanidae (horseflies), larvae	1		1			2	
Syrphidae (flowerflies).							
Unidentified larvae and pupae				4	1		1
Syrphus sp							1
Anthomyiidae, adults				1			
Asilidae (robberflies), adults			1				
Dolichopodidae (long-footed flies), adults			1				
Muscidae.							
Lucilia sp		1					
Borboridae	3						
Ephydridae, (brine flies), larvae and pupae	8	9	21	1	3	15	
Scatophagidae.							
Hydromyza confluens							1
Order HYMENOPTERA (Ants, Bees, and Wasps).							
Unidentified adults and cocoons	3		9	1	2	6	7
Tenthredinoidea (saw flies), cocoon							1
ICHNEUMONOIDEA (Parasitic wasps).							
Vipionidae.							
Protapanteles sp							1
Braconidae.							
Bassus sp							1
Meteorus sp							1
Ichneumonidae.							
Unidentified							1
Amblyteles sp				1			1
Lissonota sp							1
Metopius sp	1						
Paniscus sp						1	
Phaeogenes sp						1	1
CHALCIDOIDEA.							
Chalcididae.							
Chalcis sp							1
Spilochalcis sp	1						
SERPHOIDEA.							
Unidentified							2
Scelionidae.							
Hadronotus sp	1						

FOOD HABITS OF SHOAL-WATER DUCKS. 65

TABLE II.—*Items of animal food identified in the stomachs of the ducks treated in this bulletin and number of stomachs in which found*—Continued.

Kind of food.	Gadwall.	Baldpate.	Greenwinged teal.	Bluewinged teal.	Cinnamon teal.	Pintail.	Wood duck.
Total number of stomachs examined	417	255	653	319	41	790	413
CLASS **Insecta**—Continued.							
Order HYMENOPTERA—Continued.							
FORMICOIDEA (Ants).							
Formicidae.							
Unidentified	8	2	17	18	1	12	12
Ponera sp							2
Monomorium sp							1
Solenopsis sp							1
Crematogaster sp							18
Myrmica rubra scabrinodis							1
Formica sp							2
Camponotus herculeanus pennsylvanicus							15
Camponotus sp							4
SPHECOIDEA.							
Sphecidae.							
Diploplectron sp		1					
VESPOIDEA (Wasps).							
Unidentified			1			1	3
Vespidae.							
Polistes annularis							1
APOIDEA (Bees).							
Halictidae.							
Halictus (*Chloralictus*)							1
CLASS **Arachnida**.							
Order PSEUDOSCORPIONIDA.							
Unidentified				1			
Order ARANEIDA (Spiders).							
Unidentified	3	1	3	3			74
Argiopidae (orb-weavers).							
Tetragnatha extensa							1
Order ACARIDA (Mites).							
Unidentified			2				
Hydrachnidae (water mites)	3	1	18	22	2	11	5
SUBKINGDOM **MOLLUSCA** (MOLLUSKS).							
Unidentified	1	2	90	106	15	117	2
CLASS **Pelecypoda** (BIVALVES).							
Unidentified	3	4	3	2	2	73	8
Mytilidae.							
Mytilus edulis						1	
Veneridae.							
Anomalocardia cuneimeris						16	
Anomalocardia sp						1	
Gemma gemma						1	
Cyrenidae.							
Musculium sp							1
Sphaerium striatinum		1					
Sphaerium stamineum		1					

TABLE II.—*Items of animal food identified in the stomachs of the ducks treated in this bulletin and number of stomachs in which found*—Continued.

Kind of food.	Gadwall.	Baldpate.	Greenwinged teal.	Bluewinged teal.	Cinnamon teal.	Pintail.	Wood duck.
Total number of stomachs examined	417	255	653	319	41	790	413
CLASS **Pelecypoda**—Continued.							
Cyrenidae—Continued.							
Sphaerium sp				3			
Pisidium occidentale						1	
Pisidium sp				2			
Tellinidae.							
Tellina sp						1	
Macoma sp						1	
Angulus sp						1	
Myidae.							
Sphaenia ovoidea						3	
CLASS **Gastropoda** (UNIVALVES).							
Unidentified	5	24				71	9
Hydrobiidae.							
Bythinella monroensis						1	
Bythinella sp						2	
Hydrobia sp						2	
Acmaeidae.							
Acmaea testudinalis						1	
Tornatinidae.							
Tornatina canaliculata						5	
Pleurotomidae.							
Mangilia plicosa						7	
Amnicolidae.							
Amnicola coronata						1	
Amnicola porata				2			
Amnicola depressa							1
Amnicola cincinnatiensis		1					
Amnicola sp				3		1	
Physidae.							
Physa heterostropha			3	2		1	
Physa gyrina				1			
Physa sp	1			3		3	1
Lymnaeidae.							
Lymnaea columella				2			
Lymnaea desidiosa				1			
Lymnaea palustris				1			
Lymnaea sp						1	
Planorbis bicarinata				1			
Planorbis duryi				1			
Planorbis glabratum				2			
Planorbis parvus				4		3	
Planorbis trivolvis				4		1	
Planorbis sp	2		1	1			3
Pompholyx effusa						1	
Nassidae.							
Nassa acuta						2	
Nassa vibex						4	
Ilyanassa obsoleta						2	

FOOD HABITS OF SHOAL-WATER DUCKS. 67

TABLE II.—*Items of animal food identified in the stomachs of the ducks treated in this bulletin and number of stomachs in which found*—Continued.

Kind of food.	Gadwall.	Baldpate.	Greenwinged teal.	Bluewinged teal.	Cinnamon teal.	Pintail.	Wood duck.
Total number of stomachs examined	417	255	653	319	41	790	413
CLASS **Gastropoda**—Continued.							
Columbellidae.							
Anachis avara						5	
Anachis obesa						12	
Astyris lunata						13	
Turbonillidae.							
Turbonilla sp						1	
Pyramidellidae.							
Odostomia bisuturalis						12	
Odostomia sp						10	
Auriculidae.							
Melampus borealis	1						
Olividae.							
Olivella mutica		1					
Littorinidae.							
Littorina rudis						1	
Littorina irrorata	1						
Cerithiidae.							
Bittium nigrum						3	
Bittium varium						1	
Bittium sp						5	
Diastoma varia						11	
Cerithium sp						4	
Cerithidea tenuis						8	
Rissoidae.							
Hydrobia californica		2					
Bithinella texana	2						
Bithinella sp							1
Fluminicola nuttalliana						1	
Pyrgulopsis spinosus	1						
Assimineidae.							
Assiminea affinis						1	
Valvatidae.							
Valvata virens						1	
Valvata tricarinata		1		1			
Neritidae.							
Neritina reclivata			2			11	
Neritina virginea	25		3			40	
Neritina sp	1					7	
SUBKINGDOM **CHORDATA** (VERTEBRATES).							
CLASS **Pisces** (FISHES).							
Unidentified, teeth, scales, etc	12	2	3	3		15	5
Poeciliidae.							
Fundulus sp				2		1	
CLASS **Amphibia** (FROGS, TOADS, AND SALAMANDERS).							
Ranidae (frogs).							
Unidentified							2
Rana sp						1	

PUBLICATIONS OF THE U. S. DEPARTMENT OF AGRICULTURE RELATING TO THE FOOD HABITS OF WILD BIRDS.

AVAILABLE FOR FREE DISTRIBUTION BY THE DEPARTMENT.

The English Sparrow as a Pest. (Farmers' Bulletin 493.)
Some Common Game, Aquatic, and Rapacious Birds in Relation to Man. Farmers' Bulletin 497.)
Food of some Well-known Birds of Forest, Farm, and Garden. (Farmers' Bulletin 506.)
Some Common Birds Useful to the Farmer. (Farmers' Bulletin 630.)
Common Birds of Southeastern United States in Relation to Agriculture. (Farmers' Bulletin 755.)
The Crow in Its Relation to Agriculture. (Farmers' Bulletin 1102.)
Propagation of Wild-duck Foods. (Department Bulletin 465.)
The Crow and Its Relation to Man. (Department Bulletin 621.)
Economic Value of the Starling in the United States. (Department Bulletin 868.)

FOR SALE BY THE SUPERINTENDENT OF DOCUMENTS, GOVERNMENT PRINTING OFFICE, WASHINGTON, D. C.

Fifty Common Birds of Farm and Orchard. (Farmers' Bulletin 513, colored plates.) Price, 15 cents.
Birds in Relation to the Alfalfa Weevil. (Department Bulletin 107.) Price, 15 cents.
Eleven Important Wild-duck Foods. (Department Bulletin 205.) Price, 5 cents.
Food Habits of the Thrushes of the United States. (Department Bulletin 280.) Price, 5 cents.
Birds of Porto Rico. (Department Bulletin 326.) Price, 30 cents.
Food Habits of the Swallows. (Department Bulletin 619.) Price, 5 cents.
Food Habits of the Mallard Ducks of the United States. (Department Bulletin 720.) Price, 5 cents.
Waterfowl and Their Food Plants in Sandhill Region of Nebraska; pt. 1, Waterfowl in Nebraska; pt. 2, Wild-duck Foods of the Sandhill Region of Nebraska. (Department Bulletin 794.) Price, 15 cents.
The Relation of Sparrows to Agriculture. (Biological Survey Bulletin 15.) Price, 10 cents.
Birds of a Maryland Farm. (Biological Survey Bulletin 17.) Price, 20 cents.
The Bobwhite and Other Quails of the United States in Their Economic Relations. (Biological Survey Bulletin 21.) Price, 15 cents.
The Horned Larks and Their Relation to Agriculture. (Biological Survey Bulletin 23.) Price, 5 cents.
Food Habits of the Grosbeaks. (Biological Survey Bulletin 32.) Price, 25 cents.
Birds of California in Relation to the Fruit Industry. (Biological Survey Bulletin 34. Part 2.) Price, 40 cents.
Food of the Woodpeckers of the United States. (Biological Survey Bulletin 37.) Price, 35 cents.
Woodpeckers in Relation to Trees and Wood Products. (Biological Survey Bulletin 39.) Price, 30 cents.
Index to Papers Relating to the Food of Birds. (Biological Survey Bulletin 43.) Price, 10 cents.
Food of Our More Important Flycatchers. (Biological Survey Bulletin 44.) Price, 20 cents.
Hawks and Owls from the Standpoint of the Farmer. (Biological Survey Circular 61.) Price, 5 cents.
Destruction of the Cotton Boll Weevil by Birds in Winter. (Biological Survey Circular 64.) Price, 5 cents.

ADDITIONAL COPIES
OF THIS PUBLICATION MAY BE PROCURED FROM
THE SUPERINTENDENT OF DOCUMENTS
GOVERNMENT PRINTING OFFICE
WASHINGTON, D. C.
AT
25 CENTS PER COPY
▽

Lightning Source UK Ltd.
Milton Keynes UK
UKHW020623060119
334855UK00006B/420/P